UPLAND HABITATS

Wild, empty, bleak, desolate, hostile – a wide range of adjectives are used to describe upland habitats yet each picks up on particular features of upland environments that are a consequence of the absence of cultivation, the topography and the climate. Uplands cover almost a third of Britain's land surface, encompassing habitats as diverse as the granite tors on Dartmoor, the eroded peat plateau of Kinder Scout in Derbyshire, and the arctic-like plateau of Ben MacDui in Scotland. Yet despite obvious differences, each is recognizable as an upland environment sharing important characteristics which include the absence of human intrusion through buildings, boundaries and roads. Upland habitats are the closest we can get to wilderness in Britain.

Upland Habitats presents a comprehensive, illustrated guide to the habitat types, flora, fauna and conservation and management issues of Britain's major wilderness areas including blanket bogs, heather moors, sheep walk and deer forest, and montane and sub-montane habitats. The plants and animals found in such habitats are described, and illustrated with line drawings and photographs.

Examining the unique characteristics of scale, 'naturalness' and strong environment gradients of upland environments, this book explores a wide range of ecological processes, including: distribution patterns, adaptation, population and community dynamics. Tracing important historical events that have shaped upland habitats since the end of the last glaciation, the authors focus particularly on important conservation and management issues for the ecology and survival of British uplands as we enter the next millennium. Modern agricultural practices and economics, habitat degradation through overgrazing, commercial afforestation, recreation and persecution of wildlife, the funding of upland farming and the complex conservation-related statutes imposed to control current upland land use, are all discussed and case studies selected to highlight the issues.

Wilderness does not disappear overnight but by the time the decline in upland habitats becomes a recognized problem, it may be too late to reclaim these wild, open landscapes.

Alan H. Fielding is Senior Lecturer in the Department of Biological Sciences at Manchester Metropolitan University; **Paul F. Haworth** is a conservation consultant working for SNH, JNCC and RSPB, and a Research Associate at Manchester Metropolitan University.

HABITAT GUIDES

Series editor: **C. Philip Wheater**

Other titles in the series:

Urban Habitats
Woodland Habitats

Forthcoming titles:

Freshwater Habitats
Grassland and Heathland Habitats
Marine Habitats
Agricultural Habitats

We would like to dedicate this book to Dr Derek Ratcliffe. His outstanding contributions to so many aspects of upland ecology have been, and continue to be, an inspiration to both of us.

CONTENTS

•

PLATES

COLOUR

BLACK AND WHITE

FIGURES

•

TABLES

•

SPECIES BOXES

•

ACKNOWLEDGEMENTS

We owe a great debt to Professor Des Thompson of Scottish Natural Heritage for the enthusiastic help and advice that he gave to us during the writing of the book. The text benefited greatly from the advice of an anonymous referee, and the careful proof reading of Philippa Roberts of Earthwatch, Oxford. The editor, Dr Phil Wheater has always been constructive and cheerful in answering our many questions. Our families (Sue and Rosie and Trish, Erica and Kathryn) have been very supportive and tolerant. I (AHF) would also like to acknowledge the friendship and company of Nigel Webster, who has been subjected to many upland ecology lectures whilst we have been walking in the uplands. Dom Morgan of *Unfolding Island Images* was a great help in the processing of plates.

Just before my (AHF) father died several years ago I learned that he had spent most of the Second World War training troops in arctic warfare in the Cairngorm mountains. I suspect that my interest in the mountains comes from long forgotten childhood stories of mules, snow and cold.

SERIES INTRODUCTION

•

The British landscape is semi-natural at best, having been influenced by human activities since the Mesolithic (*circa* 10000–4500 BC). Although these influences are most obvious in urban, agricultural and forestry sites, there has been a major impact on those areas we consider to be our most natural. For example, upland moorland in northern England was covered by wild woodland during Mesolithic times, and at least some was cleared before the Bronze Age (*circa* 2000–500 BC), possibly to extend pasture land. The remnants of primaeval forests surviving today have been heavily influenced by their usage over the centuries, and subsequent management as wood-pasture and coppice. Even unimproved grassland has been grazed for hundreds of years by rabbits introduced, probably deliberately by the Normans, sometime during the twelfth century.

More recent human activity has resulted in the loss of huge areas of a wide range of habitats. Recent government statistics record a 20 per cent reduction in moorland and a 40 per cent loss of unimproved grassland between 1940 and 1970 (Brown 1992). In the forty years before 1990 we lost 95 per cent of flower-rich meadows, 60 per cent of lowland heath, 50 per cent of lowland fens and ancient woodland, and our annual loss of hedgerows is about 7000 km. There has been substantial infilling of ponds, increased levels of afforestation and freshwater pollution, and associated reductions in the populations of some species, especially rarer ones. These losses result from various impacts: habitat removal due to urban, industrial, agricultural or forestry development; extreme damage such as pollution, fire, drainage and erosion (some or all of which are due to human activities); and other types of disturbance which, although less extreme, may still eradicate vulnerable communities. All of these impacts are associated with localized extinctions of some species, and lead to the development of very different communities to those originally present. During the twentieth century over one hundred species are thought to have become extinct in Britain, including 7 per cent of dragonfly species, 5 per cent of butterfly species and 2 per cent of mammal and fish species. Knowledge of the habitats present in Britain helps us to put these impacts into context and provides a basis for conservation and management.

A habitat is a locality inhabited by living organisms. Habitats are characterized by their physical and biological properties, providing conditions and resources which enable organisms to survive, grow and reproduce. This series of guides covers the range of habitats in Britain, giving an overview of the extent, ecology, fauna, flora, conservation and management issues of specific habitat types. We separate British habitats into seven major types and many more minor divisions. However, do not be misled into thinking that the natural world is easy to place into pigeon holes. Although these are convenient divisions, it is important to recognize that there is considerable commonality between the major habitat types which form the basis of the volumes in this series. Alkali waste tips in urban areas provide similar conditions to calcareous grasslands, lowland heathland requires similar management regimes to some heather moorland, and both estuarine and lake habitats may suffer from similar problems of accretion of sediment. In contrast, within each of the habitat types discussed in individual volumes, there may be great differences: rocky and sandy shores, deciduous and coniferous woodlands, calcareous and acid grasslands are all typified by different plants and animals exposed to different environmental conditions. It is important not to become restricted in our appreciation of the similarities which exist between apparently very different habitat types and the, often great, differences between superficially similar habitats.

The series covers the whole of Britain, a large geographical range across which plant and animal communities differ, from north to south and east to west. The climate, especially in temperature range and precipitation, varies throughout Britain. The south-east tends to experience a continental type of climate with a large annual temperature range and maximal rainfall in the summer months. The west is influenced by the sea and has a more oceanic climate, with a small annual temperature range and precipitation linked to cyclonic activity. Mean annual rainfall tends to increase both from south to north and with increased elevation. Increased altitude and latitude are associated with a decrease in the length of the growing season. Such climatic variation supports different species to differing extents. For example, the small-leaved lime, a species which is thermophilic (adapted for high temperatures), is found mainly in the south and east, while the cloudberry, which requires lower temperatures, is most frequent on high moorland in the north of England and Scotland. Equivalent situations occur in animals. It is, therefore, not surprising that habitats of the same basic type (such as woodland) will differ in their composition depending upon their geographical location.

In the series we aim to provide a comprehensive approach to the examination of British habitats, whilst increasing the accessibility of such information to those who are interested in a subset of the British fauna and flora. Although the series comprises volumes covering seven broad habitat types, each text is self-contained. However, we remind the reader that the plants and animals discussed in each volume are not unique to, or even necessarily dominant in, the particular habitats but are used to illustrate important features of the habitat under consideration. The use of scientific names for organisms reduces the likelihood of confusing one species with another. However, because several groups (especially birds and to a lesser extent flowering plants) are often referred to by common names (and for brevity), we use common names where possible in the text. We have tried to use standard names, following a recent authority for each taxonomic group (see the species list for further details). Where common names are not available (or are confusing), the scientific name has been used. In all cases, species mentioned in the text are listed in alphabetical order in the species index and, together with the scientific name, in systematic order in the species list.

1

INTRODUCTION

•

WHAT ARE UPLANDS?

The British uplands, which cover almost a third of Britain's land surface, encompass a wide range of habitats ranging from places such as the granite tors of North Hessary on Dartmoor, through the eroded peat plateau of Kinder Scout in Derbyshire to the arctic-like plateau of Ben MacDui in Scotland. Despite the obvious differences between these places each is instantly recognisable as an upland environment, so what do they share in common? One important shared characteristic is the absence, or at least paucity, of human signs such as roads, buildings and boundaries. They are the closest that we can get to wilderness in Britain.

The adjectives that people use to describe upland habitats depend upon their point of view, but range through words such as open, wild, empty, bleak, desolate and hostile (MacKay 1995). Each of these adjectives picks up on particular features of upland environments that are a consequence of the absence of cultivation, the topography (shape of the land) and the climate. There is little doubt that the upland landscape is valued in Britain since nine of the twelve English and Welsh National Parks have a significant amount of upland habitat (Table 1.1). Scotland currently has no National Parks, if it did most would probably be in the uplands.

Ratcliffe (1977) defined uplands as land that is typically above the limits of enclosed farmland. Although this, and the word upland, implies an altitudinal boundary, the altitude of the land is really a surrogate for climate, since the position of the boundary between lowland and upland is related to the

Table 1.1: National Parks in England and Wales that have a significant upland component

National Park	*Area* (km^2)	*Per cent of England and Wales combined*
Dartmoor	953	0.63
Exmoor	694	0.46
Brecon Beacons	1357	0.90
Parc Cenedlaethol Eryri (Snowdonia)	2142	1.42
Peak District	1438	0.95
Yorkshire Dales	1769	1.17
North York Moors	1435	0.95
Lake District	2279	1.51
Northumberland	1049	0.69
All	**13116**	**8.68**

effect that weather has on plant growth. The upper limit of enclosed farmland is an economic threshold, above which it is unprofitable to cultivate the land. Farming close to the boundary is always marginal and changes in agricultural funding could move the boundary up or down the hill. The boundary between lowland and upland can often appear sharp, with markedly different vegetation above and below the headdyke of a farm. This sharp transition in vegetation across a farm boundary wall or fence is a consequence of the relatively intensive management that takes place on the farm. The real ecological boundary between the lowlands and the uplands is much more fuzzy but, none the less, real.

The uplands are a rewarding environment for ecologists because of their scale, 'naturalness' and strong environmental gradients. These characteristics provide excellent opportunities for the study of a wide range of ecological processes including: adaptation; population and community dynamics; and, if left alone, evolution that is not being driven by anthropogenic changes. The scale of many upland environments means that there is scope for the study of large populations that, in many cases, do not suffer the effects of anthropogenic habitat fragmentation.

Although the uplands are unlikely to suffer from the urbanisation or intensive farming experienced in the lowlands there are other threats that are a direct consequence of our actions. For example, there is habitat degradation brought about by overgrazing; commercial afforestation; recreation; persecution of wildlife and the insidious effects of acid deposition and global warming. All of these are a threat to the naturalness of the upland environment and to their value as wilderness areas. Indeed the concept of wilderness demands the existence of conscious and unconscious human attempts to interfere with nature (Bryson 1995). If wilderness can be said to have a value, then the uplands are worth a great deal to society. Wilderness does not disappear overnight, it is lost gradually as a consequence of processes that 'nibble away' at the edges; few single losses being spectacular enough to become a 'cause célèbre'. By the time that the decline in wilderness becomes recognised as a problem it is probably too late and some will make the argument that further losses do not matter because what remains has little value. It is arguable that we have no wilderness since, according to the definition of wilderness applied by the American Wilderness Act (1964), most of Britain's wildest areas would be rejected because 'man and his own works dominate the landscape' (Bryson 1995).

The book follows the same structure as the other books in this series. The second chapter provides the background information that is used to support the material in the third and fourth chapters. Choosing what should be included and excluded, or at least consigned to a rather brief coverage, was difficult. Some material, e.g. coniferous woodland, was excluded to avoid overlap with other books in the series. It would have been very easy to write a book that concentrated on birds and Scotland, largely because these have been the focus of much of the research. We have attempted to avoid this rather obvious route. We have paid particular attention to the funding of upland farming and the rather confusing set of conservation-related statutory designations. We do not think that it is possible to understand why the uplands are as they are without this knowledge. The case studies were chosen to highlight what we consider to be important aspects of the ecology of the British uplands at the start of the next millennium. There is no doubt that others would have had a different selection, but this just goes to show how diverse the uplands are.

2

THE ECOLOGY OF UPLAND HABITATS

•

WHERE ARE THE UPLANDS?

Mountains and uplands are generally found where rocks are resistant to erosion, for example over 60 per cent of the British upland and marginal upland landscapes are found where the predominant rock is acid igneous or metamorphic (Bunce *et al.* 1996). Although the rocks associated with the uplands are generally resistant to weathering most show evidence of glacial erosion from the last glaciation during the period 50000 to 15000 BP. The distributions of rock types in Britain means that uplands are largely absent from the east and south. Indeed there are no significant uplands south of a line drawn between Scarborough and Bath, while north of that line much of the land is upland or mountain (Figure 2.1). Consequently most of Britain's upland habitat is found in Scotland and Wales (Table 2.1). It is the combination of this north-western distribution, combined with Britain's position on the western edge of Europe that gives our uplands their unique character.

Figure 2.1: Distribution of upland environments within Britain and Ireland

THE IMPORTANCE OF GEOLOGY

The type of rock has profound ecological influences on, for example, the soil type and on the topography of the ground. The relationship between geology and ecology introduces a further complication because north-western Britain has an almost unrivalled diversity of rock types. Differences in the topography of the uplands are known to have marked effects on the vegetation and the breeding bird fauna (Haworth and Thompson 1990). For example, the mountains around Ben Nevis have a much poorer bird fauna than the mountains of the Cairngorms. One reason for this is the difference in landform between the two regions. Ben Nevis is a much more

Table 2.1: The extent of upland habitat in Britain expressed as the percentage of the land surface in five classes

Altitude (m)	*Type*	*England*	*Scotland*	*Wales*
<123	Agricultural land	69.4	34.0	34.0
123 – 244	Marginal agricultural	21.2	25.3	26.6
245 – 610	Hill pasture & moorland	9.1	34.3	38.2
611 – 914	Mountain	0.3	5.9	1.1
>914	High mountain	<0.1	0.5	<0.1
	Total Area (km²)	**130450**	**78829**	**20720**

Source: Based on Ratcliffe and Thompson 1988: Table 1.

rugged mountain, resulting from a collapsed volcano (Murray 1987); the Cairngorms consist of a wide expanse of granite plateau dissected by deep glacial valleys with some cliffs in the corries. Many upland boreal and arctic birds need slopes of less than 15° for breeding, hence they tend to be absent from the more rugged mountains that are usually associated with older metamorphic rocks such as gneiss and schist (e.g. NW Highlands) or younger sedimentary rocks (e.g. Snowdonia and the Lake District). A similar comparison can be made between Y Wyddfa (Snowdon, 1086 m) and Carneddau (1062 m). Y Wyddfa has many narrow ridges and steep slopes hence there is little scope for the development of upland habitat, whilst Carneddau has a small summit plateau on which woolly fringe moss *Racomitrium lanuginosum* (Species Box 2.1) heath can be found (Ratcliffe 1977).

Rocks that are resistant to weathering tend to be deficient in calcium. Since many of the

Species Box 2.1: Woolly fringe moss *Racomitrium lanuginosum*

Racomitrium lanuginosum is a moss that gets its common name from the terminating hair-like leaf tips that enable it to trap water droplets and nutrients from clouds. It grows on a variety of surfaces including bare rock and peat surfaces that are not subject to prolonged snow-lie. *Racomitrium* heath is a very common montane habitat in Northern Scotland where it forms extensive mats with other species such as the stiff sedge *Carex bigelowii*. The summit of Ben Wyvis has one of the most extensive and best developed examples of this community. It seems that the community was once more widespread but many examples, such as those in the Lake District, have been replaced by grassland communities as a consequence of sheep grazing and air pollution. *Racomitrium* is particularly susceptible to air pollution because it grows in an ombrotrophic ecosystem.

rocks of northern Britain are calcium-deficient, the resulting soils are generally acidic with a low productivity. The importance of calcium can be seen in the few upland areas where it is more plentiful, for example Ben Lawers and Y Wyddfa. In these places the flora is much richer and more similar to that found in the alpine regions of Europe. The high rainfall in the uplands exacerbates the shortage of calcium, which tends to result in leaching of nutrients from the soil. In reality the nutrients are redistributed and relatively nutrient-rich patches, called flushes, are common in the uplands. The flushes often form at a break in the slope and are characterized by much greener, more luxuriant, vegetation. These can be important food sources for the young of many wading birds such as curlew *Numenius arquata* (Species Box 2.2). Leaching also has the effect of reducing pH. The combination of high rainfall and low pH leads to the accumulation of undecomposed organic material in the form of peat. The resultant acidic, nutrient-poor soils are unable to support productive vegetation. This in turn affects the fauna, with larger numbers being found in regions with less acidic, more productive soils. For example, ptarmigan *Lagopus mutus* (Species Box 2.3) and dotterel *Charadrius morinellus* (Species Box 2.4) are found at higher densities on the richer soils of the schist mountains around Drumochter, rather than on the nutrient-poor soils on the Cairngorm plateau (Nethersole-Thompson 1973). Similarly, dunlin *Calidris alpina* (Species Box 2.5), which are found sparsely distributed on high wet peaty watersheds, reach very high densities on the rich machair regions of the Outer Hebrides.

Species Box 2.2: Curlew *Numenius arquata*

The curlew is Britain's largest wader (50 cm). It has long legs and a characteristic downward-curved bill that is used to forage for invertebrates in mud and soft soil. It is related to the smaller, and less common, whimbrel *N. phaeopus* that breeds mainly on Orkney and Shetland. The curlew breeds on sheepwalk and lowland agricultural habitats. However, unlike the redshank, whose distribution has spread further into the uplands, the curlew's distribution has spread into the lowlands during this century. The 'coorli' sound of the curlew's call, which is combined with a trilling song during courtship displays, is one of the most evocative upland noises during spring. Most curlew nest below 600 m and have well defended territories within a home range of about 8 ha. They are ground nesters and the eggs and young are potential prey for both mammalian and avian predators. In addition if there is a high density of grazers there is an additional risk of trampling. The young are fledged by mid July and the curlew begin to migrate downwards and eventually onto the coasts where they over-winter in mixed species flocks.

The enclosed grasslands along the upland fringe are an important source of food for the curlew and the improvement of these fields by draining, fertilizing and re-seeding has resulted in less usage by curlew and other upland waders. It seems probable that this has resulted in some reduction of the national populations of curlew and similar waders.

Source: Ratcliffe 1990a

Species Box 2.3: Ptarmigan *Lagopus mutus*

The ptarmigan is in the same genus as the red grouse. However, unlike the red grouse the ptarmigan is one of only three birds whose breeding is confined to the montane zone where it is widely distributed (excluding Ireland). Adult ptarmigan are herbivores, eating mainly the shoots of dwarf shrubs such as crowberry and bilberry. It has been lost from mountains, such as the Lakeland Fells, where heavy grazing has resulted in the conversion of dwarf shrub communities to grasslands.

They are ground nesters that seldom breed below 600 m, although their nests tend to be lower in the northwest where the tree line is also lower. The breeding density ranges from 1.2 to 4 pairs per hectare, with the Scottish Highlands having one of the highest densities in the world.

They are strongly territorial birds, establishing territories in February and March. Eggs are not laid until May when the weather has improved. Initially the young feed from invertebrates and develop quickly, being able to fly for short distances at 10 days old. As winter approaches their plumage becomes white and they can remain in the montane regions, scratching through the snow cover to gain access to the vegetation. If the snow is deep, flocks of ptarmigan may move temporarily down the hill.

The main predators of adult ptarmigan are the golden eagle, peregrine and fox.

Source: Ratcliffe 1990a

THE IMPORTANCE OF CLIMATE

Ecological boundaries

When there is a marked gradient of environmental conditions it is usually possible to detect zonation in the distributions of plants and animals. The classic example of this is on rocky shores. One big advantage of shore zonation is that it can be studied at a scale that is readily accessible. A similar, but less accessible, zonation also occurs on mountains. This time it is changing climate, rather than the tidal immersion–exposure cycle, that is responsible. Two major ecological boundaries can be recognized in British uplands. The first separates lowland agricultural land from the open, uncultivated uplands and it is associated with climatic limitations to plant growth, as defined by the length of the plant-growing season. Poor plant growth at higher altitudes makes it uneconomic to farm.

Species Box 2.4: Dotterel *Charadrius morinellus*

The dotterel is an arctic-alpine species which, in Britain, breeds almost entirely in the Scottish Highlands, although a few pairs can be found in southern Scotland, the Lake District and the northern Pennines. The total population is around 950 pairs.

Level summit plateaux dominated by moss and lichen heaths are much favoured for nesting. Breeding takes place at 460 m in Sutherland rising to 640 m in Rosshire and over 700 m in the central Highlands. Dotterel occupy the montane plateaux from early May until mid August. Egg laying begins in late May with a clutch of three being the norm. Dotterel differ from most birds because it is the brightly coloured male who incubates the eggs. Soon after laying the females gather together and move away, sometimes flying over to Scandinavia.

Recent research by Scottish Natural Heritage has revealed marked variation in both breeding density and productivity. A variety of factors are thought to be responsible for this variation including geology, severe weather in June and July, predation pressure and nest losses caused by sheep and deer trampling.

Source: Thompson and Whitfield 1993

The second ecological boundary is the tree line, the altitude at which trees can no longer grow. It appears that tree growth does not occur unless the mean temperature exceeds 10°C for at least two months. In addition, at higher altitudes shelter from the wind and freedom from drainage are essential for significant plant growth. As with many ecological boundaries the tree line marks a gradation of habitats. Initially trees become stunted and the forest becomes more like scrub, with an increasing frequency of juniper and willow before being completely replaced by dwarf shrub. In Britain the tree line is used to split upland habitats into two zones.

1 sub-montane (below the tree line) and
2 montane (above the tree line).

In Europe a different nomenclature (Figure 2.2) is used and Horsfield and Thompson (1996) have recently recommended adopting this terminology to avoid confusion when discussing uplands in a wider context.

As on the seashore the partitioning of the habitat into zones is somewhat complicated by local conditions. For example, factors such as the slope, aspect, soil type and climate can alter the relative and absolute altitudes of the boundaries.

One of the difficulties with attempting to describe upland climates is that data are generally unavailable. There are few accurate weather data from upland regions. Taylor (1976) and Grace and Unsworth (1988) provided detailed accounts of British upland climates. One of the most fascinating insights into mountain weather comes from an observatory on the summit of Ben Nevis (1343 m). Data were obtained between 1883–1904 (Table 2.2). Fortunately comparative data are also available for the adjacent town of Fort

Species Box 2.5: Dunlin *Calidris alpina*

The dunlin is a long-legged, long-billed wader, which together with similar birds is often categorized as a sandpiper. The dunlin is the most numerous and widespread of the sandpipers. It is much smaller than the curlew, being less than 20 cm in length. It is a ground nester in grass tussocks, typically laying four eggs in May. They are particularly associated with wetter moorland, with many pools, where their principal dipteran prey are abundant. Dunlin are often found in association with the golden plover. The association seems to offer an increased ability to detect approaching predators.

They over-winter on some estuaries in very large flocks up to 100000.

Plate 2.1: Extremes of weather – Ben More (966 m), Isle of Mull

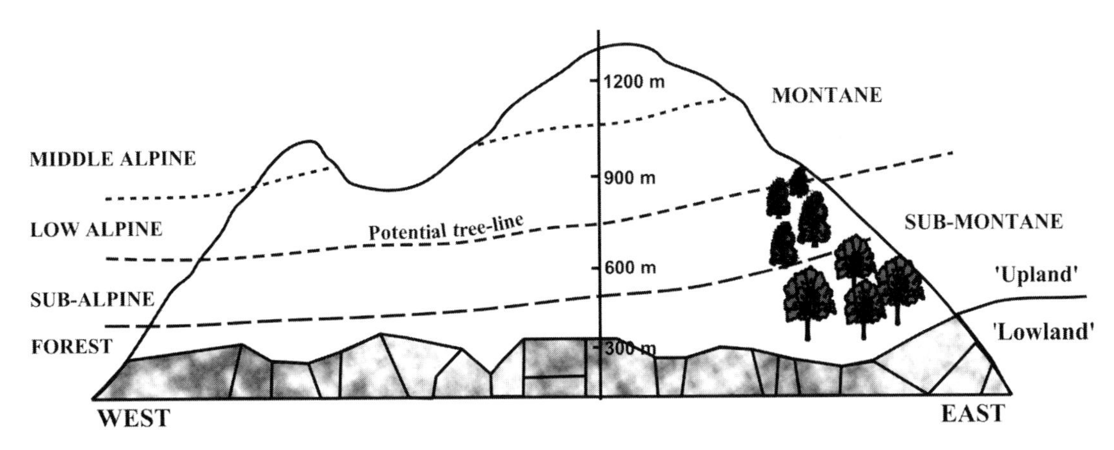

Figure 2.2: Zonation of upland habitat within Britain and Ireland: The terminology on the left is European, that on the right is the more usual British usage (Thompson, *pers comm*)

Table 2.2: The climate of Ben Nevis (BN), Fort William (FW) between 1883 and 1904 and Widdybank Fell (WF) between 1968 and 1975

	Mean Temperature °C			*Sunshine hours*			*Rainfall mm*		
	BN	*FW*	*WF*	*BN*	*FW*	*WF*	*BN*	*FW*	*WF*
Jan	–4.8	3.7	1.2	19.6	24.1	20.5	483	221	190
Feb	–4.8	3.8	–0.2	41.9	49.7	49.0	383	175	140
March	–4.3	4.7	1.3	59.8	95.2	84.0	427	178	141
April	–2.1	7.3	3.8	82.5	139.5	116.7	244	102	108
May	0.7	9.8	7.0	128.3	180.1	158.7	211	89	86
June	4.4	13.0	10.1	124.7	173.2	182.7	198	89	89
July	5.4	13.9	12.0	90.3	131.0	149.7	287	117	93
August	4.9	13.6	12.4	58.9	111.4	142.0	356	175	108
September	3.3	11.8	9.7	58.1	98.5	90.3	429	208	127
October	–0.3	8.1	7.2	46.4	75.2	77.5	376	201	107
November	–1.4	6.7	2.6	30.0	28.5	47.1	406	190	186
December	–3.5	4.5	1.8	17.1	12.7	35.3	538	287	148
	–0.17	8.4	5.7	757.6	1119.1	1151	4338	2032	1523

Sources: Pearsall 1971: Table 2, and Pigott 1978: Table 5.

William (9 m) and Widdybank Fell (510 m), which is over 250 km further south at the head of Teesdale.

The markedly different climates experienced by Fort William and the summit of Ben Nevis, which are only 6 km apart, is a graphic demonstration of why climate is so important to the flora and fauna of the uplands. The kinds of ecological change experienced during the 1300 m climb to the top of Ben Nevis are equivalent to a trek northwards of over 1000 km. For example, the mean temperature at

Lands End is 11°C; on the Shetlands it is 7.2°C; a drop of only 3.8°C.

Although the recording period differs it is possible to compare the climate of the two Scottish locations with that at Widdybank Fell. The Widdybank Fell mean temperature is intermediate, reflecting its altitude and latitude. It is, however, drier and sunnier than either Fort William or Ben Nevis. This presumably results from its position in the rain shadow of the Lake District mountains.

Oceanicity

If Britain had no upland areas the major control of climate would be the maritime factor (i.e. distance from the sea). However, not all of Britain is lowland and the uplands are not evenly dispersed across the country. Taylor (1976) described the conclusions reached by Harrison (1973) which indicated that, particularly in the west, altitude was a vastly more significant factor in temperature variation than either maritime effects or latitude. A combination of factors produces a rather unusual set of climatic conditions in upland Britain. The islands of Britain and Ireland, and particularly their uplands, lie on the western edge of the great land mass of Europe and Asia, adjacent to the Atlantic Ocean and in the path of the warm sea water of the Gulf Stream's North Atlantic Drift. This combination of circumstances produces a marked climatic gradient from west to east across Britain. One of the causes of this gradient is the fact that major cyclonic weather systems pass over Britain from the Atlantic, and the westerly orientation of the uplands shelters eastern regions (Tout 1976). It is usual to describe the climate of the west as oceanic (or maritime) whilst that of the east is more continental. The main features of an oceanic climate are the small annual variations in many climatic features such as air temperature. For example, in the SW of Ireland the difference between the warmest and coldest months of the year is only 8°C (Gilbert and Fryday 1996). Continental climates are characterized by large annual extremes in temperature caused by warming of the land when the sun is high in the sky and its subsequent cooling when the sun is low in the sky. Consequently, the differences between the two climate types arise because of differences in the relative ability of land and water to retain and conduct heat, combined with the inability of polar-maritime air masses to penetrate far inland. Although the oceanic influence is more marked in the west, its presence is still apparent further east in regions such as the Cairngorms. Conrad (1946) developed an index that can be used to demonstrate the extreme oceanic regime of NW Britain compared with the increasingly continental regime in the SE.

$$\text{Continentality Index (k)} = (1.7A/\sin(\theta + 10°C)) - 14$$

where A is the annual temperature range (°C) and θ is the latitude. Interpretations of this scale must be based on relative rather than absolute values. In Britain the values range from 1.3 at Cape Wrath in the far NW of Scotland to 12.5 at Heathrow Airport in London. Much of western Britain (e.g. Tiree 2.1, Scilly Isles 3.7) and the whole of Ireland (highest value is 6.7 at Armagh) have low values for the index. These compare with the lowest value of 0 at Thorshavn in the Faeroes. Most of Wales and northern England are sheltered from the more extreme oceanic influences because of the ameliorating effect of Ireland. This is evidenced by the distributions shown by a range of species, for example lichens (Fryday 1996). Although eastern and south-eastern locations are comparatively continental they are a long way from the extreme value of 100 found at Verkhoyansk in Siberia. As a comparison, in the USA values rise above 60 and the area with a continentality index below the UK maximum of 12 is restricted to the tip of the Florida peninsula and the Pacific coastlands (Tout 1976).

Temperature

Temperature is important because it is one of the main determinants of plant growth rates. This is reflected in the considerable geographic variation in the length and timing of the agricultural growing season across Britain. This variation is associated with the length of time that the soil temperature, 30 cm below the surface, remains above 6°C. This is ultimately dependent on three factors: altitude, slope and aspect. For example, the growing season may be 20 days longer on a south-facing slope compared with a flat site at the same location. This effect can often be seen on the opposite slopes of a valley, where enclosed grassland may be found at greater heights on the comparatively sheltered eastern side of a hill compared with western slopes. A reasonable estimate of the length of the growing season (LGS) in days can be found from the equation $LGS = 29Ta - 17$ (Ministry of Agriculture, Fisheries and Food 1976), where Ta is the mean annual air temperature (°C). The agricultural grazing season does not coincide with the growing season. This is because there is a delay while sufficient plant growth occurs to support grazing animals, and this is also altitude dependent. The delay is negligible up to 50 m (5 days), but then increases by 4 days per 100 m up to a maximum of 15 days at 300 m. The approximate length of the grazing season (LGrS) in days can be obtained from the equation $LGrS = 29.3Ta - 0.1R + 19.5$ (Ministry of Agriculture, Fisheries and Food 1976) where R is the mean annual rainfall in mm. Table 2.3 illustrates the tremendous variation in the lengths of the growing and grazing seasons across parts of England and Wales.

The altitudes at which upland and montane habitats begin are related to the rate at which temperature decreases with altitude in the British Isles. The decrease in temperature with altitude is determined by the temperature lapse rate, and Britain has some of the sharpest in the world (Manly 1970). In northern Britain the average temperature declines by about 6.7°C for each 1000 m rise. However, local lapse rates deviate from this average (Table 2.4) because of the effects of aspect, slope, proximity of the sea, weather type, time of day and season. Indeed it is possible for lapse rates to be reversed. The large lapse rate is a consequence of the oceanic climate in the west and the frequent arrival of polar-maritime air masses, which have their own large temperature lapse rates. The UK Meteorological Office has adopted a standard lapse rate of 6°C per 1000 m, which is an average of the maximum (7°C) and minimum (5°C) rates.

The tree line marks the switch from a submontane to a montane environment. A temperature lapse rate of 6.7°C per 1000 m means that the tree line begins at about 650 m. However, because the tree line is a climatic boundary its altitude will vary with location as a consequence of the increasing oceanic influence in the west and the decreasing temperatures in the north. In NW Sutherland the tree line is at 300 m and it is almost at sea-level on Orkney and Shetland. In comparison with many other temperate countries, Britain has a very low tree line. For example, in Central Norway birch scrub occurs above 1050 m and pine is found over 760 m. In Britain the tree line is almost always hypothetical because

Table 2.3: Variation in the lengths and start dates of the growing and grazing seasons of 14 English and Welsh agricultural districts. Growing season data are given as the start date and its length (days), the grazing season is defined by the number of days delay and its length

Region	*Mean height m*	*Growing season*	*Grazing season*
Fylde	28	23 March (258)	5 (199)
Coastal Kent	40	19 March (269)	5 (241)
Dorset & West Sussex	45	13 March (279)	5 (226)
Suffolk	49	26 March (248)	5 (251)
Anglesey & Lleyn	61	16 March (277)	5 (202)
Cheshire Plain	65	24 March (256)	6 (203)
Lowland Northumberland	67	4 April (237)	5 (203)
SW Cornwall	83	20 February (322)	12 (275)
Coastal Cumberland	109	3 April (237)	7 (150)
Upland Northumberland	214	21 April (200)	13 (125)
Dartmoor	236	25 March (257)	12 (125)
Upland south Pennines	287	16 April (209)	14 (119)
Upland north Pennines	315	25 April (189)	16 (111)
Cumbrian Hills	341	26 April (190)	17 (not defined)

Source: Based on data in Ministry of Agriculture, Fisheries and Food 1976

Table 2.4: Average temperature lapse rates at four locations

Region	*Lapse rate °C 1000 m^{-1}*	*Source*
Northern Pennines	6.9	Manley 1943
Aberystwyth hinterland	6.7	Smith 1950
South Wales	7.3	Oliver 1960
Ben Nevis	6.4	Table 2.2

anthropogenic effects have removed the trees. Scotland still retains a few 'real' tree lines, the most famous being at Creag Fhiaclach at the northern end of the Cairngorm mountains. At Creag Fhiaclach the tree line is at about 650 m, although scattered, stunted Scots pine can be found up to about 700 m at other localities in the Cairngorms. Pigott (1978), using evidence from existing plantations and scattered trees, estimated that the current tree line in Teesdale would be at about 600 m.

Wind and rain

The radiation, heat and moisture balances in the uplands are all affected locally by variables such as slope, aspect, vegetation cover, soil or peat cover and by local winds determined by the local landscape. There are some general trends; for example it is known that both rainfall and windspeed increase with height. For example, on the eastern coast of Scotland it rains for about 500 h each year compared with

2000 h for the NW Highlands. However, since the NW Highlands have nine times as much rain, the extra 1500 h of rain must also be more intense (Taylor 1976). The coast along the western edge of Eryri (Snowdonia) has an annual precipitation of approximately 100 cm, whilst the higher parts of Y Wyddfa have more than 400 cm of precipitation (Ratcliffe 1977). Rainfall gradients are generally higher on west facing slopes than on east facing slopes. This simple relationship is complicated by the fact that west facing slopes tend to be steeper and shorter, a factor that is sufficient to explain some of the climatic differences between west and east facing slopes. The wettest parts of the uplands are those where deep valleys, exposed to wet, westerly air streams, penetrate a long way inland, e.g. the Dovey estuary in Wales (Taylor 1976).

Atmospheric air movements cause winds but their speed is ameliorated close to the ground by the friction from the earth's surface. As the ground surface becomes rougher and higher the friction layer in the atmosphere becomes deeper and more distorted, resulting in greater windspeeds. Up to 300 m windspeeds tend to be comparable with those on the coast but over 600 m they are much stronger and more comparable with those over the open sea. Grace and Unsworth (1988) calculated a typical rate of increase in windspeed with altitude of between 6 and 9 ms^{-1} 1000 m^{-1}. The Ben Nevis data suggests an annual average of 261 gales >50 mph, this compares with only 40 at sea-level (Taylor 1976). Higher windspeeds are a major physical constraint on vegetation development, especially trees. Higher windspeeds can also have important effects because of their chill factor. This has the effect of reducing the effective temperature, possibly by over 10°C.

Both humidity and cloud cover also increase with altitude, contributing to the general wetness of the upland environment. Because atmospheric absorption of solar radiation decreases with altitude we should expect solar radiation to be higher on the summit of Ben Nevis compared to Fort William, unfortunately the additional cloud cover tends to reduce the radiation (Taylor 1976).

Snow

One consequence of decreasing temperature and increasing rainfall is that the frequency and intensity of snow increases with altitude. A number of studies suggest an average rise in snow cover of 8 days per 100 m, the rate is lowest for Devonshire moors and greatest for the Central Highlands (Taylor 1976). Uplands exposed to the generally colder, easterly winds are the snowiest, e.g. the Cairngorms, the eastern side of the northern Pennines, the Peak District and the uplands of NE Wales. Snow cover changes the surface albedo of the uplands, which means that more of the incoming solar radiation is reflected and air temperatures are lowered. Extremely low minima occur just above the snow surface. There are some places in Scotland that retain their snow for most of the year, for example the impressive Braeriach corries in the Cairngorms. These semi-permanent snowbeds can be considered to be incipient glaciers waiting for the next climatic deterioration.

Surprisingly, lying snow can be advantageous to some organisms because it insulates them from the more severe conditions above the snow. In summer time a characteristic flora can be found marking the position of snow beds. Some animals, particularly herbivores, such as ptarmigan and voles, can survive in tunnels under the snow where they can gain access to the vegetation. Indeed many organisms can only survive above 900 m because of the persistent snow cover. In more exposed locations the snow will be blown clear. It is these places that experience the most severe climatic conditions and very little can survive here. Such landscapes are marked by a range of features more associated with the high Arctic (such as boulder fields, vegetation stripes and terraces). Some of these features may be relicts from earlier, colder periods, brought about by periods of intense freezing followed by thawing and re-freezing. Lying snow can, however, create problems because it is a good 'scavenger' of atmospheric pollutants. When the snow melts quite high concentrations of undesirable chemicals may be released on to sensitive flora and fauna.

Summary

The climate of the British and Irish uplands can be characterized as a combination of low temperatures, severe wind exposure, excessive precipitation, cloud and humidity, persistent winter frost and snow cover, continual ground wetness and low evaporation. The west has a more oceanic, maritime climate, whilst the east has a less humid, continental climate with its larger annual temperature variation. Although it is possible to describe general climatic trends there are major local deviations brought about by the topography and ground conditions. Indeed, McClatchley (1996) considered that the climatic environment of British mountains was considerably more complex than previously thought.

The climatic conditions in the uplands are important because they have an overriding impact on the ecology of the uplands. One of the main factors is the shorter growing season that makes these habitats unsuitable for agriculture. However, as in the Arctic there is a significant growth of plants and invertebrates during the short summer. Summer residents, such as the birds, utilize much of this growth. At the end of the summer many organisms are forced to move, downwards and/or southwards, or dieback in preparation for the winter conditions. Very few species spend the winter at high altitudes in an active state: exceptions include birds such as the ptarmigan and the snow-bunting *Plectrophenax nivalis* (Species Box 2.6).

A SHORT HISTORY OF THE UPLANDS

Pre-historical

Although it is possible to detect some significant biogeographical trends associated with earlier times, the most important prehistoric factors affecting the uplands are a consequence of the last ice age. The period from the start of the Pleistocene epoch, approximately two million years ago, to the current Holocene epoch (Littletonian in Ireland) is associated with a very complex series of climatological events. These events reshaped much of the land surface, altered the distribution of many species and were associated with the extinction of many large mammals such as the Irish giant deer *Magaceros giganteus* and the woolly mammoth *Mammuthus primigenius*. Fossils are very important for the reconstruction of climates; knowledge of present distributions is used to infer information about historical conditions.

The last 2 million years seems to have been characterized by a cycle of warming and cooling resulting in at least six identified glacial periods, separated by interglacial and interstadial periods. Interstadials are distinguished from interglacials by their shorter duration, which prevented complete re-invasions of species that had the potential to grow under the

Species Box 2.6: Snow Bunting *Plectrophenax nivalis*

In Scotland the snow bunting is on the southern edge of its mainly arctic breeding range. The Cairngorm plateau is the best studied of the few locations where it nests at low densities in the montane zone. This work provides an insight into a species which is on the edge of its range, and presumably susceptible to the effects of climate change.

Males occupying territories sing from high boulders and occasionally other tall structures such as pylons and huts. The nests are mainly in east facing holes formed between interlocking boulders that are generally close (<30 m) to boulder fields and within 200 m of flushes and snow patches. During the summer the adults appear to have a mixed diet of mainly insects (especially crane flies) and grass seeds, although they will also use food scraps left by human visitors. When fledglings leave the nest hole they disperse to surrounding holes in boulders where they are fed a mainly insect diet by their parents. It is known that the adults make use of the torpid insects trapped on the snow fields. They also use other sites such as the edges of snow fields, flushes and boulder heaths, but seem to avoid wide expanses of grasslands that have few boulders. In late autumn, when insects are scarce, the diet switches to mainly seeds, especially where bare ground has been reseeded.

Source: Watson 1997

prevailing conditions. For example, the Upton Warren Interstadial (about 43000 BP) seems to have had mean July temperatures of about 18°C. This was sufficient to support a rich thermophilous insect fauna of over 300 species. However, trees appear to have been absent, presumably because the climatic warming was too short to allow re-colonization from the southern and western refugia (Vincent 1990).

The last ice age was the Devensian (approximately 60000–13000 BP). Most of the land south of a line between Scarborough and Bristol seems to have been ice-free, whilst north of that line the land was covered in large ice sheets. The sea-level was probably 100 m lower, meaning that Britain and Ireland were connected to Europe by a land bridge. Indeed it seems that the Thames was a tributary of the Rhine. The pollen record suggests a gradual warming in the mean July temperature from 9°C to 13°C between 13000 and 12000 BP, followed by a subsequent reduction during the Loch Lomond stadial (11000–10000 BP). Vincent (1990) describes a different scenario based on fossil ground beetles. He suggests a rapid warming to July temperatures at least as warm as the present day, followed by the almost as rapid cooling associated with the Loch Lomond stadial. The different interpretations are said to result from the inevitable inertia associated with the spread of trees, for example most require many years of growth before they are capable of producing significant quantities of seed.

There is general agreement that the last

glaciation came to an end about 10000 years ago when the climate began to become warmer. It is in the relatively short period since then that the flora and fauna of the British uplands became re-established. Although some of the upland re-colonization was from refugia further south and west, most would have been from land bridges connecting Britain to Europe. As the sea-level rose these bridges were cut and an impoverished flora and fauna was inevitable. Much of the uplands became forested, sometimes to greater altitudes than the current tree line. Presumably it would have stayed this way had it not been for the re-colonization of Britain by people. Table 2.5 summarizes the main events and the time scale.

The arrival of people

After the last ice age, and prior to the recolonization of Britain, the landscape was very different; most of the ground surface was forested. Woodland seems to have achieved its maximum extent about 8000–7000 BP (Birks 1988). During this time all of the Pennines, the North York Moors, Bodmin and much of Dartmoor seem to have had a complete cover. Other upland areas also had extensive woodland, for example the name Scafell (England's highest peak) is possibly derived from the word 'Skogafel' meaning 'wooded hill'. Caledonia, the Roman name for Scotland, is said to mean 'wooded heights'. The species compositions of these forests have been reconstructed from pollen records found in lake sediments and peat bogs. Contrary to much popular belief the predominant woodland was not coniferous. Table 2.6 shows the major types of woodland based on information in McVean and Ratcliffe (1962) and Birks (1988).

It seems that Mesolithic hunters started to clear woodland locally in England and Wales using repeat burning (Jacobi *et al.* 1976). This did not occur in the valleys; it appears to have been restricted to land above 350 m. Using pollen records Birks (1988) has recognized four major phases of woodland clearance, the first of which was probably not entirely anthropogenic.

1 3950–3700 BP NW Highlands and Skye
2 2600–2100 BP Upland Wales, England (except Lakeland Fells), Caithness

Table 2.5: The historical relationships between climate and vegetation

Years BP	*Period*	*Presumed climate*	*Dominant trees*	*Forest cover*
2000	Sub-Atlantic	Cold and wet	Alder, oak, birch, elm, beech	Declining to <10 per cent
4000	Sub-Boreal	Warm and cool	Alder, oak, elm, lime	Declining to 70 per cent
6000	Atlantic	Warm and wet	Alder, oak	100 per cent
8000	Boreal	Warm and dry	Hazel, birch, pine	100 per cent
10000	Pre-Boreal	Rapid warming	Birch	Rising to 100 per cent
12000	Loch Lomond Stadial	Cold	None–Tundra habitat	Declining to 30 per cent
13000	Allerod	Mild	Birch	Rising to 50 per cent
14000	Older Dryas	Cold	None–Tundra habitat	0 per cent

Source: Based on Vincent 1990: Figure 8.13

Table 2.6: Species composition of woodland before the first widespread human influences

Region	*Main species*
Northwest Scotland	Birch, Rowan, Hazel
Highlands	Scot's Pine, Birch, Rowan, Poplar
Cairngorms, Speyside	Scot's Pine
Southern Scotland and Argyll	Oak, Elm, Birch
Lakeland	Birch and Hazel
North York Moors	Birch, Hazel and Scot's Pine
South Pennines	Oak, Scot's Pine, Birch and Hazel
North Wales	Birch, Hazel and Scot's Pine

3 1700–1400 BP Post Roman: Lakeland Fells, Galloway, Ardnamurchan
4 400–300 BP Cairngorm and Grampian mountains

Chapter 4 has a more detailed case study of postglacial changes in the habitat of the South Pennines.

Some early travellers commented upon the loss of the last great Scottish forests. For example, Bowman (1986), a fellow of the Linnean Society of London, described a journey around Scotland during 1825. Several times he comments on the loss of trees 'We were now in the ancient Forest of Atholl; but like almost every other part of the Caledonia Sylva, it has long been despoiled of its honours . . . nothing now remains but the name.' Because the Forest of Atholl was a hunting 'forest' it had probably lost most of its tree cover many years earlier.

The last 300–400 years have seen many changes in the uplands that are a consequence of changes in agricultural practice. Some of these are discussed in greater detail later. One of the most significant historical events was the outcome of the last Jacobite Rising of Scotland in 1745. The Jacobites were defeated and a new social order developed in Scotland. Many historians consider that this eventually produced significant land use changes that can still be recognized. People were cleared from the land (Maclean and Carrel 1986) to make way for sheep and during the Victorian period the 'traditional' Scottish sporting estates became established. Although the Crofting Act of 1886 led to some redistribution of land ownership, the predominant form of land tenure in Scotland remains the large sporting estate.

Figure 2.3 summarizes some of the major historical events that have shaped the ecology of the uplands.

LAND USE IN THE UPLANDS

Introduction

Despite a general perception that upland Britain is 'natural', very little, if any, of the British wild life is unaffected by humans. The effects are at a minimum where the habitat is protected by access difficulties (for example cliffs) or environmental conditions (for example, montane habitats). English Nature (1995) estimated that 30 per cent of the British landscape was rural, the majority of which was upland. The remaining mainly lowland 70 per

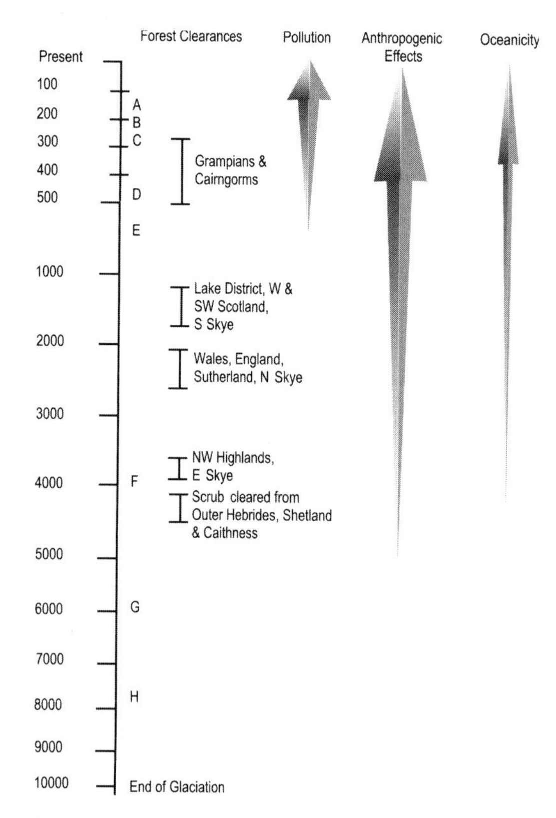

Figure 2.3: Major historical events affecting the ecology of the uplands (based on an unpublished manuscript by Thompson and Sydes)

Key:

A Start of grouse moor management
B Industrial pollution begins to change heather moor to acidic grassland in Wales and SW Scotland
C Start of intensive sheep farming
D Start of intensive burning and grazing in the Pennines
E Little 'Ice Age' creates small corrie glaciers in the Cairngorms
F Pine is extinct in much of W Scotland
G Major increase in precipitation
H Upland soils already acidic

cent is used intensively. While it is true that the upland habitats have been less affected than those in the lowlands, there are significant anthropogenic processes at work which tend to become less significant at higher altitudes.

Very crudely the effects of humans in the uplands can be summarized as grazing, burning, pollution and disturbance (Table 2.7). There is also increasing concern about the environmental impacts of windfarms. There are already major installations in southern Scotland, the west Pennines and Dartmoor and it seems likely that there will be even more in the near future.

Agriculture

Much of the moorland around the fringes of the uplands has been lost as a consequence of the expansion and intensification of agriculture and to more recent afforestation. It is estimated that 30 per cent of our uplands have been lost since 1940 due to changes in land use. This loss of habitat has resulted in the decline of some species such as the black grouse *Tetrao tetrix*. Even today upland is still being lost as a result of 'improvements' associated with better drainage, re-seeding and the application of fertilizers. Improvement is not disadvantageous to all species. The richer, more productive, fields can provide useful feeding grounds for some birds, as long as sufficient undisturbed land remains for nesting. Other agricultural improvements, such as increased stocking densities can be beneficial to carrion feeding species including the golden eagle *Aquila chysaetos* and raven *Corvus corax* (Species Box 2.7). However, improved sheep husbandry has led to fewer deaths and thus a reduction in carrion feeders.

Upland farm ownership appears to be more stable than for lowland farms. Potter *et al.* (1996) found that only 1.2 per cent of upland farms changed hands between 1984 and 1990, compared with 27 per cent of lowland farms in pastoral landscapes. This stability was also

Table 2.7: Factors influencing upland wild life

Factor	*Example effects*	*Location*
Grazing sheep	Loss of moss and dwarf shrub heath communities by conversion to grassland, which can lead to the loss of species such as the ptarmigan.	South of the Scottish Highlands and locally in the central Highlands, most of upland England and Wales
	Damage to the fragile plant communities of springs and flushes, where the lush vegetation can attract grazing sheep.	
	The nests of ground-nesting birds such as dotterel are trampled.	
Red deer	Suppression of scrub vegetation in open, high, sub-montane areas, and into the montane zone. Tree regeneration is only possible in locations where deer have no access.	Throughout Scotland, but particularly in the central Highlands
	Large herds trample the nests of ground-nesting birds when they occasionally stampede. For example, the decline of the dotterel on Moine Mhor has been linked to the large number of deer.	
Acidic deposition	Direct deposition from clouds favours the spread of grasses and sedges at the expense of globally rare mosses that may be killed in extreme conditions.	Problems are greatest in parts of the Pennines, Galloway and south of the Scottish Highlands
Global warming	Anthropogenically induced climate changes are predicted to result in an increase in average temperatures. This would have the effect of moving the upland boundaries to higher altitudes.	All of the uplands, possibly worse in the more oceanic western hills and mountains
Recreation	Spread of footpaths and associated erosion, notably in steep or boggy areas. Hill tracks used for transporting grouse and deer 'shooters' attract hill-walkers. Disturbance of some wild life is possible.	Throughout Scotland, notably on Munro mountains, other popular recreational areas close to large cities, e.g. Lake District, Peak District and Snowdonia
	Downhill ski centres attract more people to the tops, especially when easy access to high ground is maintained beyond the skiing season.	Glen Shee, Aviemore and Aonach Mor
Predators (crows & foxes)	There is a possible link between the number of walkers and the number of predators such as crows and foxes.	All popular walking areas
	Changes in the control of 'vermin' on managed grouse moors are likely to have impacts at higher altitudes.	
Alternative energy sources	The environmental impact of windfarms is difficult to assess. Their ability to generate 'clean' power may outweigh their undoubted visual impact. It is less clear how they affect upland organisms, particularly birds.	Most upland regions are potentially suitable. Changes to agricultural subsidies may make them an attractive alternative source of income for upland land owners.

Source: Based on Scottish Natural Heritage 1995: Table 9.1

Plate 2.2: Moorland fringe – pastures reclaimed for agriculture in the late nineteenth century are now slowly reverting to moorland

reflected in fewer changes in farm management practices, although it was unclear if this was a consequence of a lack of options in a difficult environment or a lack of incentives. Upland farmers are currently very concerned about their futures because of possible changes to the levels of agricultural subsidies (see Chapter 3).

With very few exceptions the upland forests have been removed, so much so that we have very few examples of natural tree lines. In Scotland the forests survived the longest but even here most were cut down by 1800. The loss of the trees resulted in the widespread development of dwarf shrub communities. These in turn were replaced by acidic grassland when they were subjected to overgrazing and burning. Thus grass-dominated sheepwalk has become widespread in many regions, particularly Wales, the Lake District and the west Pennines. In the drier east the heather *Calluna vulgaris* dominated vegetation has survived as intensively managed grouse moor.

Game management

The primary source of income for large areas of the uplands is hunting. The red grouse *Lagopus lagopus* is artificially maintained at very high densities on sporting estates by the careful management of the habitat using controlled burning and grazing regimes. Deer are well known for their ability to survive on poor land that could not support sheep. Consequently, where the soils are too poor for other uses the land is used as 'deer-forest', a mixture of poor grassland and dwarf shrub communities characterized by the almost complete absence of trees! However, in this context the term 'forest' is used in a legal, rather than an ecological, sense.

Forestry

Large areas of the uplands have been lost to forestry. Although this sounds desirable, given the earlier reduction in forest cover, it is the

Species Box 2.7: Raven *Corvus corax*

The raven is our largest corvid and throughout its range in Britain and Ireland it is associated mainly with sheep management. A combination of persecution, improved sheep husbandry and afforestation has led to declines in some areas. Breeding densities vary widely from 21 pairs per 100 km^2 in central Wales, 10–12 pairs per 100 km^2 in north Wales and 7.5 pairs per 100 km^2 in the Lake District. In Scotland high densities are also found locally within the Hebrides and Northern Isles.

Ravens are strongly attached to traditional nesting areas, indeed some nest sites can be recognized from place names such as the Gaelic Creag nam Fitheach. They build large stick nests on favoured crags or trees, laying four to six eggs in early March. Incubation lasts approximately three weeks and hatching often coincides with the availability of sheep and lamb carrion. Normally only two to three young ravens fledge from each successful nest. The young disperse from their natal areas in early autumn and often gather in large non-breeding flocks, which roost communally.

Source: Ratcliffe 1997

'wrong' type of forest consisting largely of non-native species grown in monoculture at very high densities of single aged trees. The impact of forestry is considerable and depends on the stage of afforestation. The pre-thicket period, 10–15 years after planting, leads to a decrease in species of open wet ground including dunlin, golden plover *Pluvialis apricaria* and lapwing *Vanellus vanellus*. However, this early period can lead to an increase in some important species such as the short-eared owl *Asio flammeus* and the hen harrier *Circus cyaneus*. Unfortunately as the canopy closes there are many detrimental effects. The post-thicket period, 15+ years after planting, leads to the total loss of all ground nesting birds and those encouraged by the pre-thicket period. A woodland fauna replaces the open moorland community. The next stage is the clear felling and replanting. There is some re-invasion by open ground species, but as the trees begin to grow these are lost again.

Recreation

The use of the uplands for recreation is increasing and there is considerable debate about its significance. Some species such as the merlin *Falco columbarius* and golden plover are thought to be unable to tolerate disturbance; others seem to be immune. Rock climbing has lead to the loss of some traditional cliff nesting sites. However, perhaps the main cause for concern is the increasing ease of access to wilderness areas. Many of the large sporting estates have built tracks into previously remote areas. These provide easy access for walkers and vehicles.

Skiing in Scottish ski-resorts, unlike those elsewhere, occurs above the tree line on very fragile vegetation. In addition, skiing on thin snow damages vegetation below the snow cover, and ski lift access to mountaintops puts the fragile montane ecosystem under severe pressure. Disturbance is possibly the most important impact. There is increasing pressure for more ski developments and hence the scale of disturbance will increase. Many of the ski lifts remain open in summer and allow large numbers of people access to the higher mountain regions. Apart from the obvious trampling damage there are possibly more serious effects related to the increase in birds such as crows and gulls. These birds, encouraged by discarded food, also attack the nests of vulnerable montane ground nesting species.

Military training

A significant proportion of the uplands is designated for military training, for example Mynydd Epynt in Wales and Warcop in Cumbria. Over half (13300 ha) of the highest moorland on Dartmoor is under the control of the Ministry of Defence. One of the largest ranges (22900 ha) is the Otterburn Training Area (OTA), which lies entirely within the boundary of the Northumberland National Park. The OTA is used by over 30000 soldiers each year and yet has 31 farms, 4000 ha of managed woodland, 11 SSSIs (6 per cent of the land area) and 12 sites of Nature Conservation Interest (Cross 1997). There is some inevitable disturbance to the wild life as a result of military operations but there is also the opportunity for conservation of important species and habitats. Major public conflicts arise when the military operations restrict access to walkers. Despite their negative effects on wild life it is probably true to say that a golden eagle nesting within one of these ranges would have greater protection and less disturbance than one nesting in a National Park.

More details on many of these aspects are covered in Chapters 3 and 4.

MAJOR UPLAND HABITATS AND SPECIES

British upland habitats are usually split into two broad categories, sub-montane and montane (Figure 2.2). Within each category there are a range of habitat types that are characterized by their vegetation. The term moorland is often used as a blanket term to describe much of the unenclosed, sub-montane vegetation. Unfortunately the image of what constitutes a moor is rather flexible and probably varies across the country. We will try to avoid confusion by always using the term in conjunction with a habitat descriptor, for example heather moor.

When viewed from a distance, or even better

from above, it is often apparent that upland habitats are a mosaic of different communities. In order to provide comparability between habitat descriptions it is necessary to have consistent definitions of well defined, recognizable communities. There have been several attempts to develop such definitions; for example the NCC Phase 1 Habitat Survey, the National Countryside Monitoring Scheme (Scotland) and the European based CORINE Biotopes Habitat Classification. The most recent and most comprehensive British-based scheme is the National Vegetation Classification (NVC) which recognizes over 250 relatively distinct types of plant community, of these 82 are found in the four main upland habitats (Table 2.8). Table 2.9 shows how the major heath and mire NVC communities are distributed within the British uplands.

A broader view of the relationships between the British and Irish vascular plants is given by a new biogeographical classification (Preston and Hill 1997). Their classification recognized four major biomes, each of which could be further sub-divided with respect to a longitudinal (oceanic) gradient. The three main upland groupings (Arctic-montane, Boreo-arctic Montane and Boreal-Montane) comprised 210 species (14.2 per cent of the native species) that have largely circumpolar distributions. For example, the Arctic-montane grouping consists of 79 species that are found mainly in the central Highlands and the Cairngorms. Elsewhere in Europe these species are found in Arctic, Alpine and Pyrenean habitats. Preston and Hill (1997) list 48 endemic vascular plant species, of which 25 per cent are in the largely upland groups (the Oceanic, Arctic-montane group has 3 species and the Oceanic Boreal-montane group has 11 species).

Sub-montane habitats

Heaths

This is one of the most intensively studied upland habitats (e.g. Thompson *et al.* 1995b). A heath is defined as vegetation in which dwarf shrubs such as heather, crowberry *Empetrum nigrum* (Species Box 2.8), bilberry *Vaccinium myrtillus* (Species Box 2.9) and bell heather *Erica cinerea* (Species Box 2.10) play an important structural role (Rodwell 1991a). As with most ecological classifications, the distinctions between the different types of heaths, and between heaths and grassland and heaths and mires, are often blurred. Upland heath is present over a large area of the United Kingdom (Table 2.10) and has international conservation significance because, apart from the western seaboard of Europe, it is largely absent from other countries (JNCC 1995). A shallower peat layer (<5 cm) and the absence of red grouse characterize lowland heath, which occurs below 250–300 m.

Table 2.8: NVC communities found in the uplands

Habitat	*NVC classes*	*Upland types*
Woodland	25	10
Heath	22	15
Mire (Bog)	38	30
Grassland	58	27

Source: Based on data in Thompson *et al.* 1995a.

Table 2.9: The distribution of seven heath and six mire NVC plant communities in thirteen biogeographical regions

Biographical zone	*Region*	*Heathland* H4	H8	H9	H10	H12	H16	H21	*Mire* M15	M16	M17	M18	M19	M20
Wales	A		open, frequent		light gray, frequent	light gray, extensive		light gray, frequent	open, frequent	open, frequent	open, frequent	light gray, extensive	light gray, extensive	
	B	open, frequent	open, frequent		open, frequent	light gray, extensive		open, frequent	open, frequent	open, frequent	open, frequent	light gray, extensive	light gray, extensive	light gray, extensive
England	C	light gray, frequent	open, frequent		open, frequent	light gray, extensive			open, frequent	open, frequent	light gray, frequent			
	E		open, fragments	black, extensive		open, frequent			open, frequent	open, frequent			open, fragments	open, extensive
	F		open, fragments		light gray, frequent	light gray, extensive	open, fragments	light gray, frequent	open, frequent	open, fragments	open, fragments	open, frequent	light gray, extensive	open, frequent
	G			dark gray, frequent	open, fragments	dark gray, extensive		open, frequent	open, fragments	dark gray, fragments		dark gray, frequent	dark gray, extensive	open, frequent
	H			black, extensive	open, fragments	open, frequent				black, frequent				open, frequent
Scotland	I				open, frequent	dark gray, extensive	open, fragments	open, frequent	open, frequent	open, fragments	open, frequent	dark gray, frequent	dark gray, extensive	open, fragments
	J				dark gray, extensive	dark gray, extensive		dark gray, extensive	dark gray, extensive		open, frequent	open, frequent	dark gray, extensive	
	K				open, frequent	black, extensive	black, extensive		open, frequent	black, frequent	open, fragments	open, fragments	black, extensive	open, fragments
	L				dark gray, extensive	dark gray, extensive	open, fragments	dark gray, extensive	dark gray, extensive		dark gray, extensive	open, fragments	dark gray, extensive	
	M				dark gray, extensive	open, fragments		dark gray, frequent	open, frequent		open, fragments	open, frequent	dark gray, extensive	
	N				dark gray, extensive	open, fragments		dark gray, frequent	open, frequent		open, fragments		dark gray, extensive	
Internationally significant				Y	Y	Y			Y			Y	Y	Y

Key: The size of the symbol indicates the abundance within the region:

- fragments/rare/local;
- frequent/locally abundant;
- extensive/widespread.

The shading indicates the priorities for action on heather moorland:

- black (high);
- dark gray (medium);
- light gray (low);
- open (none).

Some of the communities are internationally important.

Note: There are no upland plant communities in the SE of England

Source: The data were derived frrom Thompson *et al.* 1995a: Table 2, Figure 2 and Figure 5a

The distribution of upland heath is largely confined to regions which receive over 100 cm precipitation per year on acidic, nutrient-poor soils with shallow peat. Over most of its distribution it is maintained by rotational burning and grazing, a practice that may have started in Neolithic times, but which became more systematic after *c.*1850 (Thompson *et al.* 1995a). In the far north and west of Britain, including the outlying islands, it seems that it is able to persist without this type of management, possibly because the conditions are not conducive for tree growth. If grazing pressures are not carefully managed upland

Species Box 2.8: Crowberry *Empetrum nigrum*

The crowberry is another low growing, prostrate, evergreen shrub. It is restricted to the north and west of Britain and Ireland where it grows on dry, acid, peaty soils in rather exposed localities. The leaves are rather heather-like and, characteristically, the edges are rolled under to form tubes. Pinkish-purple flowers appear in May and June followed, in August, by brown-black berries. *E. nigrum* is a diploid species, while the related northern crowberry *E. hermaphroditium,* which is found on mountains from Iceland eastwards, is a tetraploid.

heaths easily convert to grassland, a factor that has contributed to the significant losses of this habitat that have occurred throughout Britain since the end of the Second World War. The management of conservation of heather moor is considered in greater detail in Chapter 3.

Almost all of the upland heaths are thought to have originated from the effects of grazing on the understory of cleared forests (Ratcliffe 1977). Their species composition shows considerable variation, mainly as a consequence of climatic gradients, soil moisture and anthropogenic influences including the burning cycle, stocking density and soil fertility (Thompson *et al.* 1995a). The geographical distribution of some of the major dwarf shrubs is shown in Table 2.11.

Grassland

The dwarf shrub community that is thought to have replaced the almost universal forest cover is very susceptible to grazing and burning. Both of these processes lead to increasing dominance by grasses and, in drier areas, bracken. The primary grazer is the domestic sheep *Ovis aries*. It has been present in most of the uplands, at high densities, for over 200 years, during which time there has been a relentless export of soil nutrients (in the form of lamb and wool) and intense selective grazing pressure against those plants that are susceptible to grazing. The result of this intensive grazing pressure has been the development of 'sheepwalk', a rather uniform, species-poor habitat that dominates regions such as Exmoor and Dartmoor, the Welsh mountains, the Pennines, the hills of the Lake District, the Cheviots and the southern uplands of Scotland. Because it is so extensive, and prevalent in National Parks and Areas of Outstanding National Beauty, much of the general public perceive it to be 'natural'. The importance of grazing to the ecology of the uplands is considered in more detail in Chapter 3.

Although the NVC recognizes 27 types of upland grassland there are four main types, determined by their dominant species. The first is very widespread and common in the

Species Box 2.9: Bilberry *Vaccinium myrtillus* and Cowberry *Vaccinium vitis-idaea*

The bilberry (also known as the whortleberry, bloomberry and blaeberry) is a common moorland and montane shrub where the soil is peaty. Bunce and Barr (1988) estimated that it covers 1 per cent of upland habitats. It has a low, spreading habit, rarely growing more than 50 cm above the ground. It is more exposure and shade tolerant than heather and can dominate under suitable conditions. In common with many other moorland shrubs the foliage is xeromorphic and it lacks root hairs, using mycorrhizal fungi as an alternative. Its greenish-pink flowers appear in early summer and are followed, in July, by the single, blue-black berries that are an important part of the diet of many upland animals. In winter it loses its leaves but retains the possibility of photosynthesis by virtue of its green stem.

Cowberry is a similar plant to the bilberry although it does not achieve the same height, rarely exceeding 30 cm. The flowers are pinker and the berries are red. The main difference is that cowberry is an evergreen. The origin of the scientific name is slightly confusing since it is said to relate to the 'vine of Mount Ida', a Turkish mountain that has no cowberry. It is common in Scotland and extends into the English midlands. It is generally found in rather dry, exposed situations below 400 m. In the field cowberry can be confused with the bearberry *Arctostaphylos uva-ursi*. The two species can be differentiated by their leaf characteristics, cowberry leaves are dotted with glands on the underside whereas the leaves of bearberry have conspicuous net-veins.

southern Highlands and all sheepwalk. It is characterized by three species: sheep's fescue *Festuca ovina* (Species Box 2.11), bent grasses *Agrostis canina* and *A. capillaris (tenuis)* (Species Box 2.12) [NVC communities U1, U3, U4, U5, U6, U13, U15]. As the soil becomes damper these are replaced by a moor mat grass *Nardus stricta* (Species Box 2.13) grassland [U5, U7]. This is particularly extensive in Wales, northern England and southern Scotland. As soil moisture increases a heath rush *Juncus squarrosus* (Species Box 2.14) grassland [U6] may become dominant. The final type, which the NVC recognizes as a mire or heath community, is dominated by purple moor grass *Molinia caerulea* (Species Box

Species Box 2.10: Bell heather *Erica cinerea* and Cross-leaved heather *E. tetralix*

Both of the two common *Erica* species are members of the Ericaceae family that also includes most of the moorland dwarf shrubs such as *Calluna vulgaris* and *Vaccinium* spp. Superficially *E. cinerea* and *E. tetralix* are similar but they are found in different habitats. Both species are low shrubs, seldom above 75 cm, more usually around 25 cm tall. The bell heather *E. cinerea* is a widespread plant that has leaves arranged in threes and occupies acidic and drier, peaty soils. The cross-leaved or bog heather *E. tetralix* is also widespread. It has leaves arranged in whorls of four and is found in wetter, usually peaty, soils. The flowers of the cross-leaved heather are also larger and a paler pink than in bell heather. The two species are often found in close proximity, reflecting very slight differences in surface topography and wetness.

The bell heather can be found on the drier mounds with cross-leaved heather occupying the surrounding wetter hollows.

Table 2.10: Percentage of land area classified as upland heath

Region	*Area (ha)*	*% of land area*
England & Wales	1144000	7.3
Northern Ireland	53000	3.7
Scotland	2514000	31.9

Source: Based on data in JNCC 1995

Table 2.11: The major dwarf shrubs of upland heaths and their geographical distributions

Shrub		*Main representative regions*
Ulex galli	Gorse	SW England
Calluna vulgaris	Heather	Widespread, particularly in the east
Vaccinium myrtillus	Bilberry	England, Wales and southern uplands of Scotland
V. vitis-idaea	Cowberry	Peak District and north Wales
Empetrum nigrum	Crowberry	Peak District and south Wales
Arctostaphylos uva-ursi	Bearberry	Eastern Highlands

Sources: Based on data from Ratcliffe 1977, and Rodwell 1991a

Species Box 2.11: Sheep's fescue *Festuca ovina*

Sheep's fescue is a widely distributed, hardy plant that is found from sea-level to the montane zone frequently in combination with bent grasses. Bunce and Barr (1988) estimated that it covers 2000 km^2 of the uplands. It is usually found on well-drained, nutrient-poor soils. It has significant agricultural value for upland sheep because it is very palatable and nutritious. It grows as a densely tufted perennial with bristle-like leaf blades that are rarely very tall (<25 cm). The flowers are seen from May to July on flower-heads that may be 50 cm tall. The vegetative characteristics of *F. ovina* are similar to those of the red fescue *F. rubra*, the main difference being that sheep's fescue has no rhizomes while red fescue has creeping rhizomes. This difference has potentially significant implications for the genetic structure of *Festuca* grasslands. Sheep's fescue can only spread by seed, whereas red fescue is able to spread vegetatively via the rhizomes. Berry (1977) describes studies on red fescue that suggest a single plant can cover very large areas and may be up to 1000 years old.

2.15). This is widespread and associated with wet, peaty soils. It is particularly common on Dartmoor and in Wales.

Woodland

Natural woodland would be expected to exist up to the limit of the tree line. However, natural woodland is a rather scarce habitat in Britain. Eight of the NVC woodland communities occur in the uplands (Table 2.12). There are two major environmental gradients that influence the species composition: climate and soil. Most of the woodland communities can be placed into two climatic groups: cool and wet, north western sub-montane (W9, W11, W17) and cool northern upland and montane (W18, W19, W20). Within each group there is a gradient related to the base-richness of the soil.

Species Box 2.12: Bent grasses *Agrostis* spp.

Agrostis or 'bent' grasses such as the velvet bent *A. canina* and the common bent *A. capillaris* are widespread grasses in Britain, often being found at high altitudes frequently in combination with fescue grasses. They are usually found on poor, drier acid soils and form an extensive cover of 14 per cent of the uplands, particularly where there are steep slopes (Bunce and Barr 1988). They are of less agricultural value than the fescues and if grazing is removed *Agrostis* spp. can be replaced by invading *Calluna*.

The upland oak woods (W11, W16 and W17) are believed to cover between 70000 and 100000 ha of the UK. They are an internationally important habitat that has declined in area by over 30 per cent since 1930 (JNCC 1995). The upland oak wood always has a range of other canopy species such as holly *Ilex aquifolium*, rowan *Sorbus aucuparia* (Species Box 2.16) and hazel *Corylus avellana*. Birch *Betula pubescens* (Species Box 2.17) becomes increasingly common in the canopy towards the NW of Scotland. The soil type and the amount of grazing determine the ground flora. W18 is the native pine woodland that is now found as fragmented relicts of what was once a more widespread woodland. These woodlands now occur as 77 separate blocks, totalling only 16000 ha, which is only 1 per cent of their estimated range in 4000 BP. Good examples can be found in the large Cairngorm valleys, such as Glen Derry. They are similar to the European boreal forests in which lightning-induced fire and storm damage are major controlling factors (JNCC 1995). The *Juniper – Oxalis* community, W19, may represent the climax montane scrub community, although sometimes its presence is due to the removal of other trees (Rodwell 1991b). Dwarf juniper *J. communis alpina* scrub is found growing at over 750 m on the

Species Box 2.13: Moor mat grass *Nardus stricta*

Nardus stricta is characteristic of relatively damp, base-deficient soils. *Nardus* grassland is thought to cover over 6 per cent of the British uplands (Bunce and Barr 1988). It is a tussock-forming grass that spreads vegetatively, forming a dense mat of rhizomes that few species can penetrate. Hence it tends to occur in large species-poor swards. In the montane zone its distribution is related to soil and climatic conditions, particularly snowlie. At lower altitudes the competitive ability of *Nardus* is enhanced because sheep, but not cattle, avoid eating it. This is because it has a low mineral content and becomes increasingly hard and fibrous during the year. The hard tufts of *Nardus* are often found scattered on the ground after they have been uprooted and discarded by sheep. The change from mainly cattle, to mainly sheep grazing in the uplands over the last few centuries seems to be the reason why *Nardus* grassland is now so common.

Red Cuillin mountains of Skye. The *Salix – Luzula* community, W20, is the highest woodland community in the UK and it is typically found on ungrazed, high altitude rocky slopes and ledges where there is sufficient wet, base-rich soil (Rodwell 1991b).

Mire

Mires (peat bogs) develop when soils become permanently waterlogged. Waterlogged soils tend to be oxygen deficient and this prevents the decomposition of dead plant remains, leading to a gradual accumulation of partially decomposed material that is usually called peat. There are three main types of peatland.

1 **Fens:** these develop around lake margins and other waterlogged areas where there is a supply base-rich water. The pH of fen peatlands can be above 7.
2 **Raised bogs:** these are dome-shaped bogs that can be thought of as the end of a successional process that started with a lake or pond. As they become filled with silt the water depth drops and allows the invasion

Species Box 2.14: Heath rush *Juncus squarrosus*

Rushes resemble grasses, having long narrow leaves and rather dull brown or green flowers whose parts are arranged in sixes. The heath rush *Juncus squarrosus* is found associated with boggy ground in the north and west of the country. The fibrous leaves are often flattened in a rosette on the ground, and it has a short (<40 cm), wiry flowering stem that has a cluster of terminal dark brown flowers. Pearsall (1971) looked at growth of heath rush. He found that the number of flowers, the length of the flower stalk and the number of mature capsules varied with altitude.

The fruits, which are small brown capsules about 1.5 cm long, contain numerous small seeds. The inflorescence is laid down as part of a bud in summer. It develops the following year and its length is partly a consequence of the conditions during the previous summer and in the summer during which it developed. Fruit production is more affected than growth in length, a point is reached at about 790 m above which fertile fruits are not produced, although inflorescences may be formed. This effect is said to be a result of retardation of flower and fruit development see for example some Lake District data collected in 1942.

Altitude (m)	*Flowering*
210	Complete by June
610	Not begun by end of July
790	Not complete by 1 September

In late September 1943 only one capsule per 20 plants was found on the summit of Ingleborough (723 m). Thus, despite plants being found at up to 1100 m it seems that viable seed is not produced above 800 m. However, in 1947 (a long and warm summer) viable seeds were found on the summit of Ben Wyvis (1036 m). Plants are long-lived (20+ years), thus plants above 800 m must be a result of transported seed or rare fruiting (on some mountains plants are clustered – possibly around the original parent). It is also possible that seeds are carried on the wool of sheep.

Species Box 2.15: Purple moor grass *Molinia caerulea*

Purple moor grass is a very common upland plant. Bunce and Barr (1988) estimated that it covers 6000 km^2 (10 per cent) of upland Britain. It is a widespread, tussock-forming grass that has a rather unusual characteristic. In autumn its leaves develop distinct abscission zones (Salim *et al.* 1988), similar to those on deciduous trees, and the leaves are shed forming a dense litter that can suppress other species. The plant over-winters as a swollen basal internode (<5 cm tall) with 2–4 buds from which tillers develop next spring. It is this growth pattern that is responsible for the formation of tussocks. This vegetative spread seems to be the main method of propagation since *Molinia* is poorly represented in the seed bank (Miles 1988).

In Britain *Molinia* is generally found growing as a calcifuge on wet, but not waterlogged, peaty soils. The vast *Molinia* dominated landscapes appear to have been derived from sub-montane heathland as a consequence of prolonged sheep grazing since at least 1800. Sheep are selective grazers and *Molinia* appears to be unpalatable, except for the early spring growth. In contrast cattle seem able to graze *Molinia* and under certain circumstances they may graze it sufficiently to reduce its dominance. Burning has long been used as an upland management tool, in the case of *Molinia* it seems that burning favours it over the dwarf shrubs. Hence, in combination with its litter, it is able to dominate other vegetation and produce a community that is both agriculturally undesirable and rather species poor.

of plants from the margin. Because the nutrient supply to raised bogs is from rainfall they tend to be quite acidic (pH < 4).

3 **Blanket bogs:** these consist of large carpets of peat that may extend over very large areas. They occur in regions with lots of rainfall (>1200 mm and >170 rain days $year^{-1}$), for example SW Ireland, NW Scotland and the Pennine plateaux. Under these conditions the blanket bog is the main climax community. The only supply of nutrients is from rainfall and the pH of these bogs is between 4 and 5.

Fens and raised bogs are lowland habitats. Blanket bogs are an important, and globally rare, upland habitat over much of western Britain and Ireland. The combination of waterlogging, slightly acidic rain and the organic acids released by the partial decomposition of plants creates an environment that prevents the growth of many species. The most abundant and characteristic species are the bog-mosses *Sphagnum*. Sedges, including the cotton-grasses *Eriophorum* spp., and heathers (on drier mounds) are also common plants. There are also a number of characteristic pool

Plate 2.3: Upland oak woodland – heavily grazed with little natural tree regeneration

species such as the bladderwort *Utricularia* spp. Some plants, such as the sundew *Drosera* spp. and butterwort *Pinguicula vulgaris*, are adapted to the low nutrient status of bogs by being insectivores. They supplement their nutrient supply by digesting small insects. The invertebrate fauna of blanket bogs is reasonably distinct, especially the spiders and those species found in the bog pools.

Blanket bog habitat is threatened by a number of processes (Rowell 1990). Unlike raised bogs, blanket bog is not very suitable for horticultural purposes. However, peat has been used as fuel in many upland areas. There have been some recent proposals to make greater use of some blanket bogs as a fuel resource. The more significant threats to upland blanket bogs have come from afforestation, grazing and burning. The different NVC mire communities affected by these activities are shown in Table 2.13. In Ireland only 10 per cent of its blanket bogs are thought to have any remaining conservation value. Although the Caithness and Sutherland peatlands are some of the largest and best examples of this globally rare habitat, they have suffered significant damage from afforestation (see Chapter 3), particularly during the 1980s. In 1988 half of the remaining unforested peatlands, an area of 1750 km^2, were designated as SSSIs. In other areas repeated grazing and burning have damaged the peatlands. Grazing can alter the competitive balance between species but it also has other effects resulting from trampling and nutrient enrichment from faeces and urine. It is quite difficult to burn the vegetation on peat bogs, but when it is done it tends to preferentially damage the *Sphagnum*, resulting in an increase in the abundance of shrubs such as heather.

Montane habitats

Over 90 per cent of the British montane habitat is in Scotland (JNCC 1995). There are smaller patches in Wales and on the Lakeland Fells, but in general these have suffered from more significant anthropogenic effects. Although most montane habitat is found around the summits of the highest mountains

Species Box 2.16: Rowan *Sorbus aucuparia*

The rowan or mountain ash is not an ash. The name derives from the similarity between rowan and ash leaves. Although it is widespread it is a very hardy species which is a characteristic tree of the far northwest and at higher altitudes. Some of the most northern woodlands are a mixture of birch *Betula pubescens* and rowan. Isolated rowan, often growing on crags where they are protected from grazing animals, are frequently all that remains of the higher level forests. Rowans seldom exceed 6 m and can produce large crops of bright red berries at the end of summer. The berries are an important food resource for birds. This consumption of the fruits is important for the rowan because it enables the seeds to be dispersed. The name is said to derive from the Gaelic word rudha-an (pronounced roo-ah ahn), meaning the red one.

it is possible to find similar habitats much closer to sea-level on the islands of the far north and west where the tree line is much lower. The montane habitats are important because they are mostly natural habitats and they contain many endemic species (JNCC 1995). Montane plant diversity is related to the base richness with most of the montane red data list angiosperms being associated with calcareous or base-rich soils (Table 2.14). Many of these species are very vulnerable because of their fragmented and localized small populations, for example the alpine gentian *Gentiana nivalis* (Species Box 2.18), and

Species Box 2.17: Birch *Betula* spp.

There are three species of birch that can be found in the uplands: the silver birch *Betula pendula*; the downy or hairy birch *B. pubescens* and the dwarf birch *B. nana*. Birches are very hardy species, often growing at higher latitudes and altitudes than any other tree species. They have many of the characteristics of weeds in that they do not live long (rarely over 60 years) and they produce a large amount of seed. The seeds are wind dispersed but probably not over large distances (<100 m). They do best in open conditions, for example in woodland gaps or where heather has become very 'leggy'. They are also able to grow in more acidic soils than most other trees.

The prostrate habitat of the dwarf birch is typical of many montane species. When seen in the field it may be confused with a dwarf shrub because it rarely grows more than 1 m above the ground surface. The dwarf birch, in combination with other dwarf tree species such as the dwarf willows (*Salix herbacea* and *S. reticulata*) and juniper *Juniperus communis*, seem to have been an important part of our flora at the end of the last ice age. In Britain it is now almost entirely restricted to Scotland, although it can also be found in Upper Teesdale and the Cheviots.

The hairy birch is a characteristic species of many upland woods in NW Scotland. It gets its name from the fact that young twigs are hairy. It also tends to have upright branches and bark that peel easily. The silver or common birch is less common in the uplands except in the east, but it is capable of producing the largest trees (<30 m).

Table 2.12: The main NVC woodland communities found in the British uplands

NVC code	*Name*	*Altitude range (m)*	*Soil*	*Main upland regions*
W7	*Alnus glutinosa – Fraxinus excelsior – Lysimachia nemorum* (alder & ash)	2–366	moist to very wet mineral soils	widespread, but locally, distributed in NW upland fringes
W9	*Fraxinus excelsior – Sorbus aucuparia – Mercurialis perennis* (ash & rowan)	6–335	moist brown soils derived from calcareous bedrock	N and W upland fringes
W11	*Quercus patraea – Betula pubescens – Oxalis acetosella* (sessile oak and downy birch)	15–458	moist, free draining, base poor soil (rainfall >1000 mm)	upland fringes of Wales NW England and Scotland
W16	*Quercus* spp. – *Betula* spp. – *Deschampsia flexuosa* (oaks and birches)	5–535	very acid, oligotrophic soils	Pennine fringes
W17	*Quercus patraea – Betula pubescens – Dicranum major* (sessile oak and downy birch)	12–519	very acid, shallow, fragmentary soils (rainfall >1600 mm)	widespread in W and N Britain, including Dartmoor
W18	*Pinus sylvestris – Hylocomium splendens* (Scot's pine)	16–640+	strongly leached, infertile, podzolic soils	Central and NW Scotland
W19	*Juniperus communis* spp. –*Oxalis acetosella* (juniper)	365–910+	wide range of types	E and Central Highlands
W20	*Salix lapponum – Luzula sylvatica* (willow)	630–910+	wet, mesotrophic, base-rich soils	S and Central Highlands

Note: Tree species are identified by their common names in parentheses

Source: Information is from Rodwell 1991b

the Snowdon leaf beetle *Chrysolina cerealis* which is only found in montane habitat in Snowdonia.

The montane environment is very severe. The large amounts of precipitation leach out many of the soil nutrients leaving, at best, an acidic or base-deficient soil. During the winter those places not covered by snow will experience very high wind speeds and very low temperatures. The effects of wind are very evident everywhere in the montane zone. Many plants adopt a prostrate habit to minimize wind damage; larger more luxuriant growth is only possible in the shelter of large boulders. On exposed crests, which will lack a winter snow cover, few species are able to survive. Characteristic species are the wind-resistant woolly hair moss *Racomitrium lanuginosum*, the stiff sedge *Carex bigelowii* (Species Box 2.19) and a few lichens.

In the more pristine montane areas three main community types can be recognized:

Table 2.13: Upland NVC mire communities, including those affected by management operations

Operation and NVC Code	*NVC Community*	*Main British Regions*
Created or maintained by grazing		
M10	*Carex diocia – Pinguicula vulgaris* mire	N England and Scotland
M11	*Carex demissa – Saxifraga aizoides* mire	S and central Highlands
M15	*Scirpus cespitosus – Erica tetralix* wet heath	W Highlands and SW Scotland
M19	*Calluna vulgaris – Eriophorum vaginatum* blanket mire	Pennines and central Highlands
M20	*Eriophorum vaginatum* blanket mire	S Pennines
M23	*Juncus effusus/J. acutiflorus – Galium palustre* rush pasture	SW England
M26	*Molinia caerulea – Crepis paludosa* mire	N Pennines and Lake District
Damaged by grazing		
M4	*Carex rostrata – Sphagnum recurvum* mire	NW Britain
M5	*Carex rostrata – Sphagnum squarrosum* mire	Widespread but localized
M6	*Carex echinata – Sphagnum recurvum/auriculatum* mire	Upland fringes of most of UK
M17	*Scirpus cespitosus – Sphagnum recurvum* mire	W Highlands and Western Isles
Developed under burning management		
M15	*Scirpus cespitosus – Erica tetralix* wet heath	W Highlands and SW Scotland
M19	*Calluna vulgaris – Eriophorum vaginatum* blanket mire	Pennines and central Highlands
M20	*Eriophorum vaginatum* blanket mire	S Pennines
Damaged by burning		
M17	*Scirpus cespitosus – Sphagnum recurvum* mire	
Peat Cutting		
M2	*Sphagnum cuspidatum/recurvum* bog pool community	Mid Wales, Scottish Borders
M3	*Eriophorum angustifolium* bog pool community	Eroded blanket mire, NW England
M15	*Scirpus cespitosus – Erica tetralix* wet heath	W Highlands and SW Scotland
Montane communities		
M7	*Carex curta – Sphagnum russowii* mire	Central Highlands
M8	*Carex rostrata – Sphagnum warnstorfii* mire	Central Highlands
Spring Communities		
M31	*Anthelia julacea – Sphagnum auriculatum* spring	NW Scotland
M32	*Philonotis fontana – Saxifraga stellaris* spring	Widespread
M33	*Pohlia wahlenbergii* var. *glaciallis* spring	Central and NW Highlands
M34	*Carex* demissa – Koenigia *islandica* flush	Skye
M37	*Cratoneuron commutatum – Festuca rubra* spring	Lime rich rocks, e.g. Teesdale
M38	*Cratoneuron commutatum – Carex nigra* spring	Upper Teesdale

Source: Based on Rowell 1990: Table 2

Species Box 2.18: Alpine gentian *Gentiana nivalis*

The alpine gentian is one of Britain's rarest plants. It is an annual/biennial that can only be found on two Scottish mountains, Ben Lawers and Caenlochan. Even then only about 10 colonies are known, some of which contain few plants. Because it is short-lived it depends upon seed production for survival. However, more than half of a year's crop of flowers may be consumed by grazing animals. Management of the alpine gentian populations probably requires some control of grazers, but experiments have demonstrated that populations decline if grazing is removed. This is because removal of grazing results in changes in the vegetation that allows competitors to flourish.

Source: Miller *et al.* 1994

dwarf shrub, moss-heath and grassland (there are 27 NVC montane communities, Table 2.15). If they become exposed to uncontrolled grazing all will revert to rather species-poor grassland. Some of these montane communities have a rather specialized requirement of a winter snow covering (chionophilous). Thompson and Brown (1992) thought that chionophilous and chionophobous communities were the most important contributors to montane vegetation diversity.

Dwarf shrub

There is a gradient of dwarf shrubs that can be linked to the amount of snow cover (Ratcliffe 1977). A prostrate form of *Calluna vulgaris* is found throughout much of the Highlands but is only fragmentary elsewhere. As snow cover increases more northern crowberry *Empetrum hermaphroditium* and *Vaccinium* spp. are found, eventually giving way to *Vaccinium myrtillus*.

Table 2.14: Red data montane plants

Name		*Base Rich*	*Schedule 8*
Interrupted Clubmoss	*Lycopodium annotinum*	–	–
Alpine Woodsia	*Woodsia alpina*	Y	Y
Oblong Woodsia	*W. ilvensis*	Y	Y
Scottish Scurvygrass	*Cochlearia micacea*	Y	–
Alpine Rockcress	*Arabis alpina*	Y	Y
Rock Violet	*Viola rupestris*	Y	Y
Dwarf Milkwort	*Polygala amara*	Y	–
Alpine Catchfly	*Lychnis alpina*	–	Y
Edmonston's Mouse-ear	*Cerastium archcum* ssp. *edmonstonii*	–	–
Scottish Pearlwort	*Sagina x normaniana*	Y	–
Snow Pearlwort	*S. intermedia*	Y	–
Bog Sandwort	*Minuartia stricta*	Y	Y
Alpine Sandwort	*M. rubella*	Y	–
Arctic Sandwort	*Arenaria norvegica* ssp. *norvegica*	Y	Y
Alpine Milk-vetch	*Astragalus alpinus*	Y	–
Yellow Oxytropis	*Oxytropis campestris*	Y	–
Shrubby Cinquefoil	*Potentilla fruticosa*	Y	–
Marsh Saxifrage	*Saxifraga hirulus*	Y	–
Drooping Saxifrage	*S. cernua*	Y	Y
Highland Saxifrage	*S. rivularis*	Y	–
Tufted Saxifrage	*S. cespitosa*	Y	Y
Irish Saxifrage	*S. rosacea*	Y	–
Iceland Purslane	*Koenigia islandica*	–	–
Diapensia	*Diapensia lapponica*	–	Y
Spring Gentian	*Gentiana verna*	Y	Y
Alpine Gentian	*G. nivalis*	Y	Y
Rock Speedwell	*Veronica fruticans*	Y	–
Alpine Bartsia	*Bartsia alpina*	Y	–
Purple Coltsfoot	*Homogyne alpina*	Y	Y
Norwegian Cudweed	*Gnaphalium norvegicum*	–	–
Alpine Fleabane	*Erigeron borealis*	Y	Y
Norwegian Mugwort	*Artemisia norvegica*	–	–
Alpine Sow-thistle	*Cicerbita alpina*	Y	Y
Snowdon Lily	*Lloydia serotina*	Y	Y
False Sedge	*Kobresia simpliciuscula*	Y	–
Mountain Bog Sedge	*Carex rariflora*	–	–
Small Jet Sedge	*C. trofusca*	Y	–
Hare's-foot Sedge	*C. lachenalii*	–	–
Bristle Sedge	*C. microglochin*	Y	–
Wavy Meadow Grass	*Poa flexuosa*	–	–

Key:

Base rich – indicates those species associated with base-rich and calcareous soils

Schedule 8 – indicates those species that require special protection

Source: Based on Ratcliffe 1991: Table 1

Species Box 2.19: Stiff sedge *Carex bigelowii*

The stiff sedge is a perennial plant that is usually restricted to higher altitude montane habitat, often in conjunction with *Racomitrium*. The sedge appears to be able to gain a competitive advantage over *Racomitrium* if they are subjected to grazing (Jondottir 1991).

The sedge spreads vegetatively using creeping rhizomes. Vegetative reproduction seems to be the norm. Flowering plants tend to be rare, even when grown under controlled, optimum conditions. Presumably this is an adaptation to the severe montane environmental conditions. The stems, which are obviously triangular in section, are between 5 and 30 cm long. One way of distinguishing *C. bigelowii* from the common sedge *C. nigra* is the presence of red-purplish-brown scales on the rhizomes (Jeremy and Tutin 1968). It appears that the restriction of *C. bigelowii* to montane habitats is not purely physiological since it can be grown at lower altitudes in a heated greenhouse (Pigott 1978). Presumably it is biological interactions that restrict the distribution of *C. bigelowii*, possibly because other species are unable to tolerate the severe climatic and edaphic conditions in the montane zone.

Moss heath

A community dominated by the moss *Racomitrium lanuginosum* and the sedge *Carex bigelowii* is said to be the single most extensive near-natural terrestrial community in Britain (Thompson and Whitfield 1993). It is a common summit community from north Wales northwards. There are some rarer bryophyte dominated communities associated with snow beds in the north and west Highlands: *Rhytidiadelphus loreus* and *Dicranum starkei* are found where there are late and very

Table 2.15: NVC communities found in montane environments

NVC Class	*Community*
H12	*Calluna vulgaris – Vaccinium myrtillus* heath
H13	*Calluna vulgaris – Cladonia arbuscula* heath
H14	*Calluna vulgaris – Racomitrium lanuginosum* heath
H15	*Calluna vulgaris – Juniperus communis spp. nana* heath
H16	*Calluna vulgaris – Arctostaphylos uva-ursi* heath
H17	*Calluna vulgaris – Arctostaphylos alpinus* heath
H18	*Vaccinium myrtillus – Deschampsia flexuosa* heath
H19	*Vaccinium myrtillus – Cladonia arbuscula* heath
H20	*Vaccinium myrtillus – Racomitrium lanuginosum* heath
H21	*Calluna vulgaris – Vaccinium myrtillus – Sphagnum capillifolium* heath
H22	*Vaccinium myrtillus – Rubus chamaemorus* heath
M7	*Carex curta – Sphagnum russowii* mire
M8	*Carex rostrata – Sphagnum warnstorfii* mire
M33	*Pohlia wahlenbergii* var. *glaciallis* spring
U7	*Nardus stricta – Carex bigelowii* grass heath
U8	*Carex bigelowii – Polytrichum alpinum* sedge heath
U9	*Juncus trifidus – Racomitrium lanuginosum* rush heath
U10	*Carex bigelowii – Racomitrium lanuginosum* moss heath
U11	*Polytrichum sexangulare – Kiaeria starkei* snow-bed
U12	*Salix herbacea – Racomitrium heterostichum* snow-bed
U13	*Deschampsia cespitosa – Galium saxatile* grassland
U14	*Alchemilla alpina – Sibbaldia procumbens* dwarf-herb community
U15	*Saxifraga aizoides – Alchemilla glabra* banks
CG12	*Festuca ovina – Alchemilla alpina – Silene acualis* dwarf-herb heath
CG13	*Dryas octopetala – Carex flacca* heath
CG14	*Dryas octopetala – Silene acualis* ledge community
W20	*Salix lapponum – Luzula sylvatica* scrub

Source: Modified from JNCC 1995

late lying snow beds respectively (Ratcliffe 1977).

Grassland

A number of grasses and sedges can form montane communities. The species-poor *Festuca vivipara* (Species Box 2.20) – *Agrostis canina* grassland is thought to be derived from heathland in the southern Highlands. *Nardus stricta* and *Juncus squarrosus* communities are associated with prolonged snow cover in the Highlands. *Carex bigelowii* and *Juncus trifidus* are found on the high plateaux of the Highlands. *Deschampsia cespitosa* communities are restricted to steep slopes in the Highlands where they are thought to be a consequence of grazing (Ratcliffe 1977).

Species Box 2.20: Viviparous fescue *Festuca vivipara*

In many respects viviparous fescue is a very similar grass to sheep's fescue. Apart from its restriction to montane habitats, the main difference is an apparent adaptation to living at high altitudes that is shared by some other montane and arctic plants such as the alpine meadow grass *Poa alpina*. Viviparous fescue does not produce seeds, instead of flowers small plantlets are produced from small buds. These drop off the parent and then begin to grow independently, close to the parent. Presumably the adaptation is necessary because both wind and insect pollination are risky strategies in the montane zone and the growing season is too short for seed development.

Aquatic habitats

Upland aquatic environments are almost all acidic and oligotrophic. The acidity and oligotrophy are both a consequence of the calcium-deficient rocks that are common in upland regions. Some lowland, naturally oligotrophic, waters have become eutrophic because of nutrient run-off from anthropogenic activities. This has not happened to many upland streams and lakes because there is little intensive agriculture and many of the streams are in drinking water catchments. Some upland streams around the edges of the Pennines did experience serious pollution during the early part of the century. This was because the fast moving water provided power for the early parts of the industrial revolution, pollution of the water often being an inevitable consequence.

The flora and fauna of many upland streams is often poor because of the effects of frequent torrential flows following heavy rainfall. The flora is often restricted to bryophytes and epilithic algae. Although the invertebrate fauna seems to be limited by nutrient deficiency and climate there are some specialist species such as the flatworm *Crenobia alpina*, the mayfly *Ameletus inopinatus* and the stonefly *Protonemura montana*. One interesting adaptation is that some of the upland adult stoneflies have shorter wings than the same species at lower altitudes (Ratcliffe 1977). The dipper *Cinclus cinclus*, grey wagtail *Motacilla cinerea* and common sandpiper *Actitis hypoleucos* are all associated with upland streams where they feed on the invertebrates such as stonefly larvae. The dipper was lost from many streams during the 1980s. This loss was thought to be a consequence of water acidification brought about by conifer afforestation. Continuing problems with water acidity still limit the distribution of the dipper.

Upland lakes and lochs are almost always characterized by very clear, possibly peat-tinged, acidic water. Many of the highest lochs seem to have very sparse zooplankton. A July survey of the relatively large Loch Einich (503 m) in the Cairngorms failed to detect any zooplankton, even though this loch supports charr that generally feed from the zooplankton (Ratcliffe 1977).

UPLAND FAUNA

Invertebrates

In comparison with the plants and birds the upland invertebrate fauna, particularly that in the montane zone, has been poorly studied. There are many gaps in our knowledge about the ecology and distribution of even some common species.

Climate, soil conditions, vegetation and management practices such as burning and grazing determine the upland invertebrate fauna. The effects of these processes on the fauna of six sub-montane habitats are summarized in Table 2.16 and the role of grazing is considered in the next chapter. Ratcliffe (1977) recognized 53 upland Lepidoptera, of which the majority (48) are moths. Ratcliffe (1977) suggests that the low number of butterflies is a consequence of the reduced sunshine, whilst some of the moths have adapted to the cold night time conditions by becoming active during the warmer days. There appears to be an altitude dependent, inverse relationship between the standing crop of Lepidoptera and Diptera, with Diptera increasing with altitude (Coulson 1988). Much of this increase in the Diptera is a consequence of changes in the abundance of Tipulidae (crane flies); there is a strong, positive correlation between Tipulid density and increasing altitude and rainfall. There are said to be 61 species of spider found in upland environments (Ratcliffe 1977): comprising 17 montane, 30 sub-montane and 14 widespread species. Odonata (dragon- and damselflies) are rare breeders in upland streams. This is not too surprising since they are at the northern edge of their range in Britain. Ratcliffe (1977) said that there is only one truly upland species, *Cordulegaster boltoni*, although another ten breed at the lower edge of the sub-montane zone in mire pools, ponds and lakes.

Although the invertebrate fauna of the montane zone is less studied than that of the sub-montane habitats some general trends have been recognized. For example, the diversity of insects on Scottish mountains decreases with altitude (Rotheray and Horsfield 1995), although it is unclear if this is due to worsening climatic conditions or a simple species–area relationship. Inevitably, the dependence of invertebrates on the environment for temperature regulation places certain constraints on montane species. It seems that a range of adaptations have developed. For example, some are smaller and darker than lowland relatives, others take longer to develop. The population dynamics of the cranefly *Tipula montana* have been studied by Scottish Natural Heritage (Thompson and Whitfield 1993) because of its importance as prey for the dotterel. The cranefly apparently exhibits a biannual emergence pattern that is synchronized within, but not between, individual mountains. It appears that dotterel know, in advance of the emergence, which mountains will experience a cranefly swarm.

Some of the invertebrates found in the montane zone will have been transported there by air movements from lower altitudes. In a study of insects marooned on snow patches Ashmole *et al.* (1983) found that only 10 of the 130 species identified were montane. On the snowfields the lowland species are torpid, and are easy prey for some specialist montane

Species Box 2.21: Rush Moth *Coleophora alticollella*

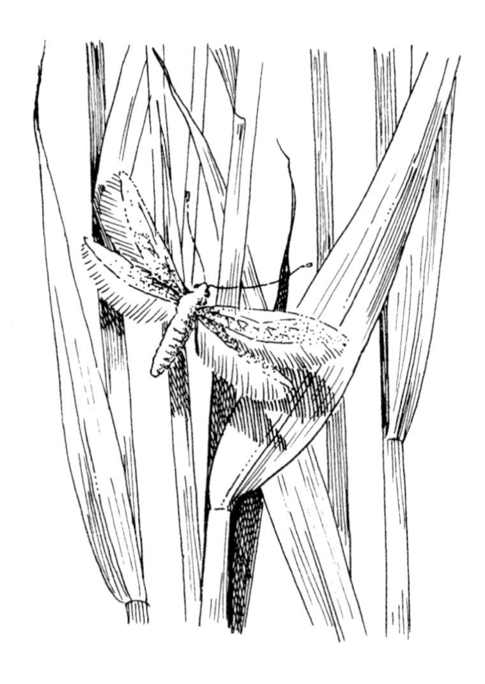

The rush moth is a very common, but easily overlooked, invertebrate that can consume a significant proportion of the annual seed production. The larvae are only 3 mm long and the adult is small and rather dull. Despite these unpromising characteristics Coulson (1978) thinks that it provides a good example of the problems faced by animals living in the uplands, namely:

1 the distribution of its food supply;
2 the distribution and abundance of predators and parasites;
3 the effects of the climate, particularly those related to low temperatures;
4 the impact of plant-feeders on the vegetation.

The phenology of the rush moth's life cycle is given below.

June	Adults emerge from the soil and lay eggs in the flowers of *Juncus*, particularly *J. squarrosus* and less often *J. effusus* (the common rush).
July	The eggs hatch and the larvae bore into developing seeds, which they then eat.
August–September	The last two instars can be seen as silk cases that protrude from the inflorescences.
October	The caterpillars drop from the plant and over-winter in the soil litter.
Spring	The caterpillar now pupates.

As Pearsall (1971) showed, the seed production of the heath rush is related to altitude, at higher altitudes seeds may not be set each year. Above 750 m the heath rush very rarely sets seed, above 600 m seed set is irregular whilst below 500 m seed is set most years. Inevitably this has consequences for the rush moth. In years when little or no seed is set at higher altitudes by the heath rush the rush moth is found only at lower altitudes (<500 m) unless the common rush is available as an alternative host (up to 600 m). It seems that common rush seed production is less affected by decreased temperatures. This alternative host is important because it provides a refuge from which the heath rush can be reinfected when temperatures increase. The ecology of the rush moth is complicated by the effects of at least seven ichneumid parasites. These parasitic wasps are rarely found above 460 m so they are only able to control the rush moth at the lower end of its distribution. The consequence of this difference in altitudinal ranges is that the rush moth probably has its greatest effect on heath rush seed production between 460 m and 500 m, the zone in which seed is set most years and the moth is free of its parasites.

Source: Coulson 1978

Species Box 2.22: An upland reed beetle *Plateumaris discolor*

Plateumaris discolor belongs to a group of reed beetles that are predominantly found in the lowlands. The adult is quite large (2–3 cm), often with a bright metallic colour. There are over 20 species of reed beetle, most are in the genus *Donacia*. All of the reed beetles spend the first part of their lives as larvae and pupae attached to the roots or rhizomes of aquatic plants. *Plateumaris discolor* is an upland species (up to at least 600 m) that is usually found attached to the roots of cotton grasses *Eriophorum* spp. It appears to be very similar to a lowland species *Plateumaris sericea*. The main problem faced by the larvae of *P. discolor* is that the roots of the upland cotton grasses are in water or mud that is almost devoid of oxygen. The beetle has evolved an interesting set of adaptations that enable it to survive. All of the plants that it uses have large intercellular air spaces. The beetle has two anal spines, attached to its tracheal system, that are used to penetrate the plant's air spaces and gain access to an oxygen supply.

The pupa does not have these spines, instead, prior to constructing the cocoon, it gains access to an air supply by eating a hole through to the root cortex. It seems that metamorphosis to the adult beetle is completed by September but the adult does not emerge until the following June.

Source: Pearsall 1971

invertebrate predators and birds such as the Snow Bunting.

Fish

It seems that few fish managed to spread further than the Scottish lowlands at the end of the last glaciation (Ratcliffe 1977). Unfortunately the present distribution of fish in British inland waters is often complicated by introductions. However, it is safe to say that there are no montane fish since, despite attempts to introduce them, none have been found in the few lakes above 850 m. Species such as the charr *Salvelinus alpinus* and vendace *Coregonus albula* seem to be glacial relicts that have survived in a few deep lakes in the north and west. Since most upland streams and lakes are oligotrophic the most common fish are salmonids including the salmon *Salmo salar* and the brown trout *Salmo trutta*.

Amphibians and reptiles

The following account is based on Ratcliffe (1977). There are no obligate upland species and those that move up from the lowlands appear to be thinly distributed. Because they are ectotherms they can experience the same cold weather problems as the invertebrates. None the less the common lizard *Lacerta vivipara* is widespread to at least 760 m. The adder *Vipera berus* is also widespread in the uplands below 600 m, but is absent from, or at least scarce in, some large areas such as Snowdonia and the Pennines. They seem to be particularly

Table 2.16: The invertebrate fauna of a range of sub-montane habitats

Habitat	*Notes*
Base-rich *Festuca ovina* grassland	A widespread, but localized, habitat which supports molluscs (snails and slugs), ants, earthworms and many insect larvae. This habitat is often recognizable by the presence of moles *Talpa europea*. Worms may form the majority of the standing crop.
Festuca – Agrostis grassland on base-deficient soils	Snails, but not slugs, are rare. Both ants and earthworms are also uncommon. Dipteran larvae, particularly leather jackets *Tipula* spp. are more common. There are a variety of small moths, e.g. *Procus strigilis* that feed, as larvae, on grass roots.
Nardus and *Molinia* grasslands on damp and very base-deficient soils	The soil fauna is very poor, but some moths may be seen frequently, e.g. antler moth *Cerapterix graminis*, fox moth *Macrothylacia rubi*. The larva of the rush moth *Coleophoa alticollella* [Species Box 2.21] lives in the fruits of the *Juncus squarrosus*.
Sphagnum mire communities	A very rich invertebrate fauna particularly associated with the pools. During summer the midges (Chironomids) and black flies (Simulium) will be obvious to anyone working in this habitat. Midge larvae can survive in water that is oxygen-deficient, whilst the black fly larvae need well-oxygenated running water. The invertebrate fauna tends to be similar to that in the Scandinavian sub-arctic regions.
Eriophorum mire communities	This appears to support a very poor invertebrate fauna, a factor reflected in the very low density of insect eating birds such as the meadow pipit. However, it is also possible that this reflects the very short period, during June, when insects become abundant. There are some specialized species such as the cotton-sedge moth Haworth's minor *Celaena haworthii*; a daddy-long-legs *Tipula subnodicornus* and a rather specialized beetle *Plateumaris discolor* [Species Box 2.22].
Calluna heathland	Usher and Thompson (1993) reviewed the invertebrate community. They found that the species composition could be explained by two environmental gradients: soil wetness and time since last burned. This is a comparatively diverse fauna. Usher (1992) found 15.3% (54 species) and 20.4% (127 species) of British ground beetle and spider species on the North York Moors. Usher and Thompson (1993) think that the explanation for this high diversity is the spatial and temporal variation in the heathland habitat. There is one invertebrate, the heather beetle *Lochmaea suturalis*, that can be both economically and ecologically significant [Species Box 2.23].

associated with *Molinia* grassland and heather moors where their prey is abundant. In areas where the adder is common it can become prey to some of the larger raptors such as buzzards. Frogs *Rana temporaria* occur at surprisingly high altitudes (>900 m) and their spawn can be found in some of the highest Cairngorm lochs. As with some of the insects the summer is too short to complete their development and some tadpoles over-winter, prior to metamorphosing into adults during their second year.

Mammals

Britain has a rather poor mammalian fauna so it is perhaps not too surprising that there are relatively few upland mammals. The current British mammalian fauna has lost nine species

Species Box 2.23: Heather beetle *Lochmaea suturalis*

As the name suggests this beetle is associated with heather. On the wetter heather moors it has the potential to be a major pest because it feeds on the bark and leaves of young heather plants. The eggs of *Lochmaea* have been found on *Sphagnum*, but it is not thought to be an obligate relationship. There does, however, seem to be a requirement for high relative humidity for successful incubation. The larvae feed mainly during July and August, they then burrow into the soil and hibernate. The small (<7 mm), dark brown-black, adults emerge the following spring. The recommended control measures, which include summer burning (to kill the larvae) and draining (to remove suitable conditions for egg incubation), have detrimental effects on other species. There are predators of *Lochmaea* such as the ladybird *Coccinella hieroglyphica*, but they seem unable to prevent the outbreak of *Lochmaea* plagues. Usher and Thompson (1993) did not think that there were sufficient data to be certain about the importance of *Lochmaea* as a pest of upland heaths.

Source: Pearsall 1971

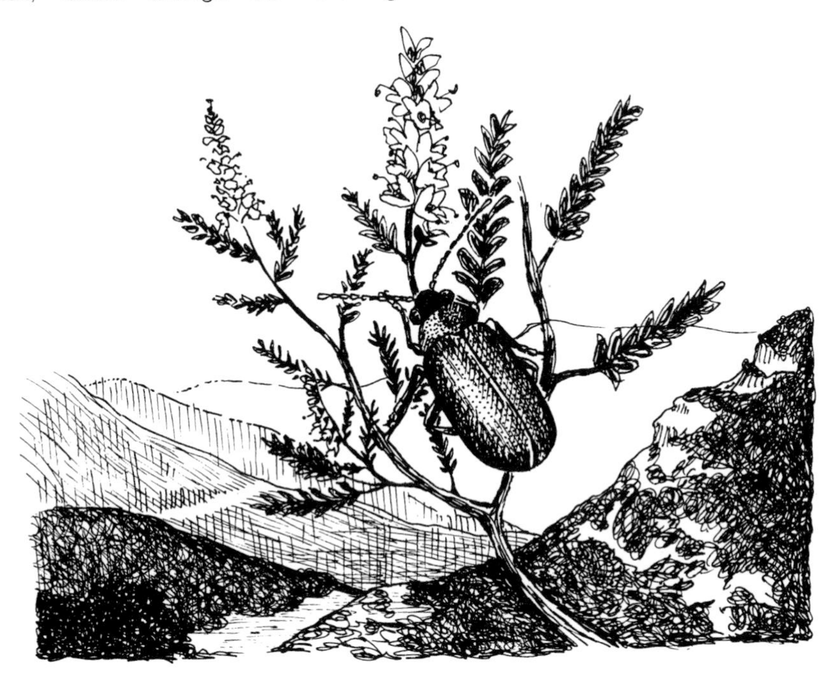

(steppe pika *Ochotona pusilla*, beaver *Castor fiber*, the northern vole *Microtus oeconomus*, wolf *Canis lupus*, brown bear *Ursus arctos*, horse *Equus ferus*, wild boar *Sus scrofa*, elk *Alces alces* and aurochs (wild cattle) *Bos primigenius*) since the end of the last ice age. Many of these species would have been found in the northern and upland habitats.

The most obvious and ecologically significant mammals in the uplands are the large herbivores, particularly the domestic sheep and, in the Scottish Highlands, the red deer *Cervus elaphus*. Feral goats *Capra hircus* can also be locally important. The mountain hare *Lepus timidus* (Species Box 2.24) is the largest mammal associated almost exclusively with

Species Box 2.24: Mountain hare *Lepus timidus*

The mountain hare differs from the brown hare *L. capensis* by virtue of its smaller size and greyer coat. In winter the coat colour becomes mostly white. There is a larger Irish form *L. timidus hibernicus* that has been introduced into some British locations (e.g. Mull). The mountain hare is almost entirely restricted to Ireland and Scotland. The exceptions are two introduced populations on the Isle of Man and the English Peak District.

The optimum habitat is said to be heather, which is their winter main food, although grasses can be important in spring and summer. It seems that dwarf shrubs are less important to the Irish race. In Ireland mountain hares are found at all altitudes and in a range of habitats, presumably because the brown hare is absent. Dietary studies of the Irish mountain hare (Wolfe *et al.* 1996) found that their diet was almost entirely grass (94 per cent), a greater proportion than is found in the diet of the brown hare. This led Wolfe *et al.* (1996) to conclude that when both species occur the mountain hare is restricted, by competition, to higher altitudes where they are forced to increase the proportion of dwarf shrub in their diet.

They are difficult animals to see since they are largely nocturnal. They have home ranges that are maintained by agonistic interactions (except between females). The size of the home range varies between season and sexes, with densities varying from 3 to 300 km^2, depending on soil productivity. Mountain hares are one of the preferred foods of the golden eagle.

Source: Hewson (1991)

the uplands. Rabbits *Oryctolagus cuniculus* and brown hares *Lepus capensis* are seldom found above 300 m, even where the mountain hare is absent (Hewson 1991). Many of the smaller mammals including the field vole *Microtus agrestis* and bank vole *Clethrionmys glareolus*, moles *Talpa europea*, wood mice *Apodemus sylvaticus* and common shrews *Sorex araneus* often extend their ranges to over 800 m and above. Most of the mammalian predators, particularly the fox *Vulpes vulpes*, weasel and stoat *Mustela erminea*, exploit the potential avian and mammalian prey of the uplands. The field vole is probably the most abundant upland mammal (except on the islands which it never colonized), particularly where the vegetation is lush enough to provide some cover for their movements. Pine marten *Martes martes* and wild cats *Felis silvestris* are mainly restricted to the woodland habitats.

Birds

The upland avian fauna is probably the best recorded and most intensively studied animal group (e.g. Ratcliffe 1990a). Ratcliffe (1990b) considered that while only 67 of the 230 British breeding birds use the uplands, many of them are protected because of their national and international conservation significance (Table 2.17). Although much of the British

Plate 1: Golden eagle – an icon of the uplands

Plate 2: Red deer stag – Britain's largest surviving mammal

Plate 3: Short eared owl on its nest – a crepuscular forager of voles on moorlands, bogs and young conifer plantations

Plate 4: A patchwork quilt of well managed grouse moor in the south Pennines

Plate 2.4: Raven – an important component of the upland predatory and scavenging bird assemblage

upland bird fauna characterizes the northern or montane element of the European bird fauna, very few species spend their winters in the uplands. The majority of the upland birds exploit the large food resource that is available during the spring and summer months, to enable them to breed. Those that remain in the uplands throughout the year tend to be raptors, scavengers or vegetarian. The invertebrate feeders need to move away from the uplands, often migrating further south to track their prey. The ptarmigan is the outstanding upland bird, often remaining on or close to the tops throughout the winter. Some of the species listed in Table 2.17 are also found in the lowlands, often around the more rugged coasts; prime examples include the raven and the golden eagle.

Ratcliffe (1990b) suggests that there are six groups of birds for whom the uplands are important:

1 Southern representatives of distinctive boreal, sub-arctic and arctic communities: for example the montane habitats supports ptarmigan, dotterel and snow bunting; the greenshank, golden plover, dunlin and arctic skua on the northern flows of Caithness and Sutherland;
2 Uniquely British bird communities: especially the birds of heather moor such as red grouse, merlin, hen harrier, short-eared owl, golden plover and ring ouzel;
3 Species with disjunct or restricted global distributions: for example the great skua and the twite;
4 High breeding densities and/or large populations: for example ptarmigan, red grouse, peregrine and golden eagle;
5 Unusual forms of northern species: for example the adaptation of greenshank to the treeless flow country instead of its more usual open tundra forest habitat;
6 Migratory species

The red grouse, golden eagle and hen harrier are considered in more detail in the next two chapters.

Table 2.17: Principal breeding birds of the British uplands

Species	*HM*	*IIB*	*RDB*	*ECB*	*Schedule 1*
Montane					
Ptarmigan	—	—	—	—	—
Golden Plover	Y	—	Y	Y	—
Dotterel	Y	—	Y	Y	Y
Purple Sandpiper	Y	—	Y	Y*	Y
Dunlin	—	—	—	Y*	—
Snowy Owl	Y	—	Y	Y	Y
Snow Bunting	Y	—	Y	—	Y
Sub Montane					
Black-throated Diver	Y	Y	Y	Y	Y
Red-throated Diver	Y	—	Y	Y	Y
Common Scoter	Y	—	Y	—	Y
White-tailed Eagle	Y	—	Y	Y	Y
Golden Eagle	Y	Y	Y	Y	Y
Buzzard	—	—	Candidate	—	—
Red Kite	Y	—	Y	Y	Y
Hen Harrier	Y	—	Y	Y	Y
Peregrine	Y	Y	Y	Y	Y
Merlin	Y	—	Y	Y	Y
Kestrel	Y	—	—	—	—
Red Grouse	Y	Y	Y	—	—
Black Grouse	Y	—	Y	—	—
Curlew	—	Y	—	—	—
Whimbrel	Y	—	—	Y*	Y
Wood Sandpiper	Y	—	Y	Y	Y
Greenshank	Y	—	Y	Y*	Y
Temminck's Stint	—	—	Y	Y*	Y
Great Skua	Y	Y	Y	—	—
Arctic Skua	Y	—	Candidate	—	—
Short-eared Owl	Y	—	Candidate	Y	—
Raven	—	—	Candidate	—	—
Dipper	—	—	Candidate	—	—
Ring Ouzel	—	—	Candidate	—	—
Meadow Pipit	—	—	—	—	—
Twite	—	—	Y	—	—

Key:

HM – birds that have some association with heather moor

IIB – birds that breed in the UK in internationally important numbers

RDB – Red data book species

ECB – birds listed in the EC Birds Directive as requiring special conservation measures

* indicates migratory species

Schedule 1 – Species on Schedule 1 of the 1981 Wildlife and Countryside Act.

Sources: Based on Ratcliffe 1990b: Table 1, and Brown and Bainbridge 1995: Table 5.1

SUMMARY

The ecology of some upland groups and habitats, such as birds, flowering plants and *Calluna* dominated moorland, is relatively well understood. The main processes that determine the species composition of upland communities are the climate, the geology and the past and present management systems. The role of some of these is described in more detail in the next two chapters. Other groups, such as the invertebrates, bryophytes and lichens are less well studied, although detailed information is available for a few 'hot-spots' such as the base-rich Ben Lawers. More detailed studies in less accessible and fashionable locations would provide valuable information on the ecological processes that are operating in the uplands.

3

MANAGEMENT AND CONSERVATION

•

In this chapter we aim to introduce some of the processes and issues that are important to the current upland environment. Although they are presented in six separate sections there are some inevitable overlaps. For example, grazing pressures are inextricably linked to agricultural subsidies.

PROTECTION AND SUBSIDIES

Much of the habitat of upland Britain is affected in some way by statutory conservation designations or by entitlement to agricultural subsidy (Table 3.1). Often these can be in conflict. Most of the recent nature conservation and agricultural policies originated within the European Union, for example EU conservation policies rely heavily on the Birds Directive (79/409/EEC) and on the Habitat Directive (92/43/EEC). Agricultural subsidies, such as Hill Livestock Compensatory Allowance (HLCA) payments, can create conflict because they are production-linked incentives that support farming in economically unsustainable regions. Inevitably, these payments have resulted in some overstocking and overgrazing. However, the EU intends to move HLCA payments away from production towards a more environmentally focused scheme.

Scotland has almost 16000 hill farmers and in 1994 the average hill farmer received £24800 in subsidies, including LFA and HLCA payments. Despite this their annual income was estimated to be around £10000 per annum (UK average income was £16500). At least 5000 of these farmers had an income that was below the official poverty line. The main problem is with the specialist sheep farmers in the remoter parts of Scotland. The HLCA subsidies were cut by 28 per cent before the 1996 payments and it has been suggested that they should be reduced further or removed altogether. It is unclear what effect these changes in subsidy will have on upland farmers and how any changes will affect the upland habitat.

There is little doubt that patterns of land ownership and agricultural subsidies have important consequences for upland conservation. In addition, throughout most of Britain land use is almost always 'estate-centred', with little or no attempt at coordination over a broader scale. When this is combined with the frequency with which some estates change ownership it becomes almost impossible to consider implementing management practices where it may take decades to bring about the desired environmental improvements. Almost inevitably this results in a confrontational approach to the implementation of national conservation priorities that may be very difficult to police in wilderness areas.

GRAZING

Upland plants are often subjected to significant grazing pressure; indeed much of the

Table 3.1: A summary of the major conservation designations and agricultural support schemes that apply in the British uplands

Designation	*Example*	*Implementation and Background*	*Notes*
World Heritage Status (WHS)	No upland sites	World Heritage Convention 1972. These are sites of international importance. A high level of legal protection is demanded.	The Cairngorms have been suggested but existing levels of management and protection are not thought to be appropriate.
National Park (NP)	Lake District	National Parks and Access to the Countryside Act 1949. NPs were established to protect landscapes and provide recreation opportunities. They are probably better defined as 'protected landscapes'. They are not state owned, rather there is usually a wide mix of land ownership and land use. Most of the control of developments is via planning consents.	The designation has not yet been extended to Scotland. Approximately 1.3 million ha is designated as National Park.
Area of Outstanding National Beauty (AONB)	Forest of Bowland, Lancashire	National Parks and Access to the Countryside Act 1949. The aim is to protect AONB from inappropriate development whilst recognizing the needs of forestry, agriculture and other rural industries. An Advisory Committee that includes representatives from the local authority, landowners and community interests manages each AONB. Although there is no statutory protection the designation is expected to contribute to outcome of planning decisions.	Many people think that the 39 AONB are a second division of National Park. Approximately 2.0 million ha is designated as AONB.
National Nature Reserve (NNR)	Beinn Eighe	National Parks and Access to the Countryside Act 1949. The 118 NNRs were established to protect the most important ecological and geological habitats. Most are owned by the relevant national conservation agency or approved bodies such as Wild life Trusts. Most have a resident warden.	They receive the same level of protection as SSSIs. Approximately 71000 ha are designated as NNR.
Site of Special Scientific Interest (SSSI)	Malham–Arncliffe	National Parks and Access to the Countryside Act 1949, amended by the Wildlife and Countryside Act 1981. The government is obliged to notify any land that it considers to be of special floral, faunal, geographical or physiological interest. The criteria used to judge potential sites include: naturalness, rarity, size, recorded history, fragility and diversity. Landowners must notify the national agency if they wish to undertake a Potentially Damaging Operation	Most of the current 6500 SSSIs are owned privately. They cover approximately 7 per cent of the English land surface and 10 per cent of that in Wales.

Table 3.1: continued

Designation	*Example*	*Implementation and Background*	*Notes*
		(PDO) on a SSSI. The national agency will negotiate with the landowner and may elect to compensate the owner for not undertaking the PDO.	
Special Protection Area (SPA)	Moor House Cumbria	1979 EU Birds Directive All EU countries are committed to taking requisite measures to protect suitable habitat for all wild birds. There is a list indicating conservation priority. In particular SPAs are expected to protect areas of international importance for rare or vulnerable species (e.g. merlin and golden eagle) and for migratory species (e.g. waders and wildfowl). SPAs should already be SSSIs but benefit from extra legal protection. In any relevant planning application conservation is expected to override economic or recreational concerns.	There are 87 proposed SPA for the United Kingdom (excluding Scotland).
Special Area of Conservation (SAC)	Ben Alder (sub-arctic willow scrub)	Special Areas of Conservation (SACs) were proposed under the Habitats Directive and receive the same level of protection as SPAs for birds. Together the two series are known as Natura 2000. Each EU member state must compile a list of all sites containing the 168 habitat types and 632 species that are thought to require conservation. The final list of SACs is due by June 1998 (see Anon. 1995).	SACs are selected from existing SSSIs. Currently there are 178 proposed SACs (excluding Scotland).
EU Birds Directive	Merlin	European Union 1979. Also known as the Directive on the Conservation of Wild Birds. This directive lists birds that are deemed to be of special conservation concern within the EU. Member states are expected to safeguard the habitats of these species through the designation of SPAs.	Later broadened to other groups by Directive 92/43/EEC.
EU Habitats Directive	Caledonian Pine Forest	European Union 1992. Also known as the Directive on the Conservation of Natural Habitats and of Wild Fauna and Flora. This directive requires member states to identify habitats and species of 'community interest'. Action must be taken to protect these species and habitats through the designation of SACs and Species Action Plans (see Hopkins 1995).	Complements the earlier EU Birds Directive.

Table 3.1: continued

Designation	*Example*	*Implementation and Background*	*Notes*
Environmentally Sensitive Area (ESA)	Island of Mull	Agriculture Act 1986. Many valued landscape features such as hedges, ditches and field barns are the product of traditional farming practices. ESA designation is intended to protect these features by preventing depopulation and agricultural intensification. Farmers enter a voluntary scheme and agree to participate in a 10 year management agreement. They receive payments as compensation for the potential income that they lose by continuing with traditional, less intensive farming methods.	By 1996 43 ESAs were established that covered 14 per cent (34000 km^2) of Britain.
Less Favoured Area (LFA)	Snowdonia NP	European Union 1975. LFAs are those regions where natural features (e.g. altitude and latitude) place constraints on agricultural production. The LFA policy was adopted to support farmers for environmental purposes. Sheep annual payments are currently about £22.50 per ewe.	53 per cent of Britain is designated as LFA, including almost all of Scotland (90 per cent) and most of Wales (75 per cent).
Hill Livestock Compensatory Allowance (HLCA)		European Union 1975. The LFA Directive enables member states to additionally compensate farmers for natural handicaps. HLCAs are annual headage payments for sheep and cattle that are kept for breeding purposes. The HLCA sheep stocking limit is 6 ewes ha^{-1}. The government recognizes different levels of disadvantage and adjusts the payments to take account of this. For example, in 1998 payments ranged from £2.65 per ewe to £5.75 per ewe for certain hardy breeds in Scotland.	HLCA payments are an important part of an upland farmer's income.

current upland habitat is a direct consequence of persistent and heavy grazing. However, as Thompson *et al.* (1987) note, the term overgrazing does not mean the same thing to ecologists and agriculturists. Ecologists recognize overgrazing as a process that alters vegetation from its natural state to a semi-natural or artificial condition. The most obvious upland grazers are the large mammals, in particular sheep and red deer. Less obvious are the smaller mammals including rabbit, hare and vole, and the large number of invertebrates. The term grazing is usually applied to grass and herb consumption, whilst the term browsing is applied to the consumption of woody plants including heather and trees.

Plate 3.1: Fence line – heavy sheep grazing to the left, ungrazed deep heather to the right

Sheep grazing in the uplands has a long history and is one of the major determinants of the upland environment. Indeed the treeless landscapes of regions such as the Welsh uplands and the Lake District fells are now thought by many visitors to be 'natural'. In reality these are unstable habitats that are maintained by heavy grazing and/or burning. If grazing was reduced the vegetation would be expected to revert back to woodland, although unavailability of seed may make this a very slow process (Hope *et al.* 1996). Recently sheep stocking densities have risen, almost certainly as a consequence of the headage payments that encourage more sheep. Highest densities (>1000 km^{-2}) are in the Welsh uplands, Exmoor, the Pennines and the Lake District, and the eastern part of southern Scotland.

Coincidentally with the increase in sheep there has been a massive rise in the Scottish red deer population. During the early 1800s there were very few red deer, but as more sporting estates were established their numbers began to increase. In 1960 the total population was thought to be around 155000, since then numbers have doubled to around 300000 (Clutton-Brock and Albon 1992). Almost everyone, including the sporting estates, now agrees that there are too many red deer and the habitat is suffering from their grazing.

Control of grazing is considered to be one of the main upland management options. It is therefore important that we understand the mechanics and effects of grazing in the uplands. Unfortunately predicting the effects that grazers will have on vegetation is not always simple because there is a complex set of interacting factors that determine how the plants will respond to grazing. Some of the main elements of upland grazing systems are described in Table 3.2.

As ecologists we should be able to predict how ecological systems will respond to change, for example how changes to the type and intensity of grazers will affect the vegetation. The ability to predict successfully depends on our understanding of the system. Information about the relationships between plants and grazers can be obtained from observational studies (we record various statistics but we do not intervene) and experimental studies (sites where we manipulate the grazers and/or the vegetation). The most common grazing experiments involve manipu-

Table 3.2: Components of an upland grazing system

	Comments
Grazer characteristic	
Red Deer	Selective grazers that are more prone to browse than are sheep and cattle.
Cattle	Unselective grazers that use their tongues to tear or strip vegetation, i.e. they do not bite. Damage from trampling or rubbing can be significant.
Sheep	Selective grazers that use teeth on their lower jaw to bite through the vegetation to produce a quite closely cropped sward. Upland breeds appear more prone to browsing than lowland breeds.
Rabbits	Selective grazers that produce a very closely cropped sward that, because of its height, can be unusable by other grazers. Rabbit grazed vegetation is often characterized by the presence of tall, unpalatable plants.
Hares	It is often thought that mountain hares have a higher proportion of shrubs in their diet than the lowland brown hare. However, the work of Wolfe *et al.* (1996) suggests that grass is the preferred food for both species.
Voles	The short-tailed or field vole is probably the most numerous upland mammal. Voles can reach high densities, but they will not use short vegetation. They eat young shoots and seeds and may bite through long stems at ground level.
Invertebrates	The importance of invertebrate grazing is probably grossly underestimated because they are not very obvious unless they reach pest proportions (e.g. the heather beetle). Grazing of plant roots is particularly difficult to quantify but it must be severe in some habitats such as those that have large tipulid populations.
Plant features	
Tree	Apical meristems make them susceptible to damage from mammalian browsing, if this is too severe saplings will probably not survive. If browsing is not too severe the plant form will be affected. Topiary is an extreme, anthropogenic example of this. If the tree becomes large enough the upper branches may escape browsing. At higher altitudes climatic factors tend to keep trees small so they may lack this opportunity. Excessive seed consumption will restrict woodland expansion opportunities.
Dwarf shrub	Apical meristems again make them susceptible to damage from browsing. However, unlike trees, they never have the opportunity to grow large enough to escape mammalian browsing. Severe browsing results in death.
Grasses	Grasses have basal meristems. Therefore, removal of the leaf tip does not prevent continued cell division and leaf expansion. This is the main reason why they are better able to tolerate grazing. Grasses differ in their palatability. Consequently grazing can alter the competitive balance between different grass species.
Productivity	The ability of a plant to tolerate grazing or browsing depends on its productivity and the amount of the production that is removed by grazers (utilization). There is an approximate relationship between these two factors such that the more productive species can tolerate a greater percentage utilization.
Phenology	Plants do not all grow and produce leaf material and seeds in unison. There are some marked differences in vegetative and flowering phenology, which changes the relative attractiveness of species during the course of the year.
Community structure	The effect that grazers have on the structure of plant communities depends on a number of factors including the ratio of grasses to shrubs and their spatial relationships.

Table 3.2: continued

	Comments
Anti-grazing strategies	Plants use a variety of anti-grazing mechanisms that can affect their palatability; for example fibrous plants (e.g. soft rush), rough cuticles (e.g. moor mat grass) and aromatic compounds (e.g. thyme). One less obvious mechanism involves alkaloid toxins synthesized by fungal endophytes. A detailed study of red fescue on St Kilda found that levels of infection and the amounts of toxin were related to the sheep grazing pressure. Similar relationships were not found on two other Hebridean islands (Bazely *et al.* 1997).
Indirect Effects	
Trampling	Larger grazers can damage vegetation by trampling; this is particularly true of dwarf shrubs. If the soil is wet footprints can break the surface cover and leave bare patches that provide germination microsites, which may encourage greater plant species diversity.
Dunging	The structure of dung differs greatly between cattle and other grazers. Cattle dung can have immediate effects on vegetation because plants may now be in the dark and experiencing nutrient overload! All dung and urine can increase soil nutrient status locally. If latrines are used (e.g. rabbits) there can be very marked local effects. More generally nutrient enrichment from dung and urine is thought to disadvantage mosses and favour grasses.
Sward height	Small mammals are more numerous where the vegetation is tall. This is important for raptors such as owls and kestrels.
Invertebrate density and diversity	Grazing has direct and indirect effects on the invertebrate communities in the vegetation. For example maximum overall biodiversity occurs when heather moorland has a mosaic of wet and dry areas combined with a large amount of structural diversity. Too much grazing (>2 ewes ha^{-1}) creates an open canopy which favours the more common carabid species at the expense of more localized and less common species (Gardner *et al.* 1997). It is possible that grazing can affect the food supply of many birds. For example, it is now known that the highest breeding success of black grouse occurs on lightly grazed moors, irrespective of whether a game keeper is used to control potential predators (Baines 1996). Indeed it is now thought that overgrazing is one of the main reasons for the decline of the black grouse.
Burning (Muirburn)	Burning is often used as a management tool to improve the quality of upland grazing by increasing the soil nutrient status or by removing unpalatable species. The effects of a muirburn can be quite variable depending, for example, on the state of the heather and the fire temperature. Philips (1981) highlighted the following effects. • If heather is young (<15 cm tall) burning tends to favour fire-resistant species such as the undesirable moor mat grass and purple moor grass. • If the heather is mature (between 20 and 30 cm) heather can quickly regrow from axillary buds. • If the heather is degenerate it is unlikely to regenerate from buds following a fire. Any new growth is likely to be from seed and such regeneration may be suppressed by grazing. • If a burn is close to bracken the bracken rhizomes will escape the fire's effects and it can regrow quickly, possibly shading out other species. • If the vegetation is damp the effects of the fire will be quite different than for dry vegetation. Controlled muirburn for grazing improvement is usually carried out between November and April when the soil and vegetation are usually damp. If there is a summer fire it is possible for the temperature to rise sufficiently to ignite the peat. This type of fire can cause permanent changes to the vegetation.

Plate 3.2: Highland cow – an increasingly familiar component of grazing in the Scottish Highlands

lating grazing densities, often using fences. At the extreme an exclosure is used to stop grazing by the target animal. Even then it is important to understand that sheep-proof fencing would not stop red deer, rabbits, voles or invertebrates from grazing within the exclosure.

One of the most intensively studied grazing systems is that on heather moorland (for example Grant and Armstrong 1993). This system has been studied because loss of heather cover is a conservation concern (Thompson *et al.* 1993) and it is also a problem for sporting estates where a significant portion of their income is related to the number of red grouse. Heather moorland is usually thought to be a semi-natural vegetation, which is maintained by burning and grazing. Too much grazing or burning moves the vegetation towards grassland, whereas too little tends to enable scrub and woodland to develop. The general dynamics between a range of vegetation types is now quite well understood (Miles 1988). In general severe grazing (>2 sheep equivalents ha^{-1}) produces a rather species-poor grassland that is likely to be dominated by unpalatable species such as moor mat grass. A more diverse habitat, that incorporates a significant cover of dwarf shrubs and trees, requires less grazing. It is difficult to be too precise about grazing intensity since burning regimes and the soil type can compound the effects of grazing.

Sheep and red deer seem to have similar effects on individual heather plants, but their impact on the vegetation may differ because their spatial distributions differ as a consequence of their different social structures (Stewart and Hester 1998). The effects that sheep and deer have on heather communities is not simple or easy to predict because they appear to be related to the spatial pattern of heather fragments within a grassland matrix. The size and spatial arrangement of heather fragments seems to be important because both species appear preferentially to graze heather that is adjacent to palatable grass species. In addition if the heather becomes highly fragmented deer and sheep movements through the patches can cause physical damage which speeds up the decline (Stewart and Hester 1998). Because heather grazing is generally restricted to the perimeters of fragments this will further affect the spatial fragmentation.

Grant and Armstrong (1993) described the feeding value and sheep feeding preferences for a range of common moorland species (Table 3.3). A feeding value above 8 is required for the maintenance of a non-lactating ewe, above 18 is needed for lactating ewes. A heather-only diet would not provide sufficient resources for a ewe. Thus all sheep require a diet that includes a significant proportion of grass. There are definite plant preferences in the approximate order: herbs; broad-leaved grasses; fine-leaved grasses; shrubs. Despite the low nutritional status and unattractiveness of heather it can be an important component of a grazer's diet. As the proportion of heather declines it becomes increasingly likely that sheep will graze it. Also heather can provide an important supply of resources in winter when grass growth has stopped, hence its relative preference rating increases.

Anderson and Radford (1994) described the results of an eight-year study of grazing on the Kinder Estate, which is an important part of the Peak District National Park. Since the National Trust became the owner in 1982 they have attempted to reduce sheep grazing so that the vegetation could revert to a dwarf shrub community from its present largely grassland vegetation.

The initial grazing density was 2.5 ewes ha^{-1}, which was too high for the maintenance of a dwarf shrub community. In some regions, particularly those offering shelter from the wind and rain, densities were much higher. Part of the problem faced by the National Trust was that most of the sheep were 'trespassers' from surrounding farms. Consequently it was impossible to control sheep density with any precision. The solution was to regularly gather the sheep and return them to their owners. The response of the vegetation to the reduced grazing was assessed by recording data from permanent transects. Rates of change were low, particularly on the steeper slopes. The main species showed a variety of responses, although all of the selectively grazed species showed an increase they did so in different ways. The responses of the five main plants are summarized in Table 3.4.

Most mammalian grazers are known to browse trees. At best this inhibits growth, at worst it kills the tree. In many of the Scottish uplands browsing of tree saplings by red deer is thought to be responsible for preventing forest regeneration. If this is perceived as a problem there are two solutions. First, deer densities can be reduced, almost inevitably by

Table 3.3: Feeding values (FV, digestible organic matter intake g kg^{-1} live weight) and ranked seasonal feeding preferences for five upland plants and communities
Lower preference numbers indicate greater preference

Vegetation	*FV*	*Spring*	*Summer*	*Autumn*	*Winter*
bent–fescue grassland	17–27	3	1	2	3
moor mat grass	13–23	3	4	4	4
purple moor grass	16–25	4	4	5	5
cotton grass	8–14	3	5	5	4
heather	3–10	5	5	5	4

Source: Based on data in Grant and Armstrong 1993

Table 3.4: Responses of five species to reduced grazing pressure

Species	*Comments*
Wavy hair grass	Spread vegetatively to form larger plants and new plants established from seed.
Bilberry	Almost all of the increase in this species was vegetative. No new plants were found suggesting that colonization by bilberry is unlikely if grazing is removed.
Heather	Heather was generally absent at the start of the trial so there was little opportunity for vegetative spread. However, on all but the steepest slopes there was good seedling establishment.
Moor mat grass	This generally unpalatable species benefited from reduced competition when there was excessive grazing; hence its failure to expand following removal of grazers was not too surprising. There was some expansion on the steeper slopes where there was a lot of bare ground that enabled moor mat grass seedlings to develop free from competition.

Source: Based on information in Anderson and Radford 1994

culling. Second, fences can be used to establish exclosures. Fencing is an effective, albeit expensive, solution that has some undesirable side effects:

- Regenerating trees tend to be the same age; this can produce woodland that lacks structural diversity.
- Deer-proof fencing needs to be very tall and this can cause problems for low flying birds such as black grouse and capercaille *Tetrao urogallus* (Baines and Summers 1997).
- If exclosures are established, but there is no culling, there will be an increase in grazing pressure outside the exclosure.

Stewart and Hester (1998) have reviewed the literature on the relationships between tree regeneration and deer density. There appears to be some disagreement about absolute deer densities and tree regeneration, although most agree that winter densities should be below 5 km^{-2} and less if broad-leaved regeneration is desired. This is partly because pine saplings

Plate 3.3: Feral goats – a locally important browser (Isle of Mull)

are less sensitive to browsing than are broad-leaved species such as rowan and birch. A further complication is that the deer use different altitudes at different times of the year, consequently susceptibility to browsing may be season-dependent.

It is beyond dispute that grazing is one of the most important processes affecting the ecology of the uplands. In general the current number of grazers is too large for the continuation of natural vegetation. Unfortunately controlling grazing in the uplands is difficult. The type and density of larger grazers is controlled mainly by economic factors such as agricultural subsidies. Revision of these policies is possible only within the wider context of food production throughout the European Union.

GROUSE MOOR

Land managed primarily for red grouse shooting is an extensive and important anthropogenic habitat that is found throughout much of northern England and Scotland. These moors cover around 1–1.5 million ha of upland Britain and are characterized by a gentle rolling topography and a prevalence of heather, which is the staple diet of the red grouse. Grouse moors show an easterly distribution because heather tends to grow better in the drier east. Many moors also contain flat waterlogged areas covered by blanket bog that contribute greatly to their overall biodiversity.

Grouse shooting began in the late eighteenth century, becoming most fashionable after the 1860s with the introduction of the more efficient breech-loading shotgun. At this time a highly specific form of land management developed involving the employment of gamekeepers. Their role was to rid estates of all so called 'vermin' suspected of predating red grouse and to burn the heather moorland on a carefully planned rotation that maximized both food and nesting cover.

Red grouse are monogamous, territorial birds. Cocks establish territories in the autumn, defending them vigorously until the next summer, although harsh winter weather may force temporary abandonment (Jenkins *et al.* 1967). Adult grouse are browsers that feed

Plate 3.4: Muirburn – on a south Pennine grouse moor

mainly on heather, although other plants such as cottongrass and bilberry are frequently taken in spring and summer. The nest is usually placed in tall mature heather with the first eggs being laid in late April. Soon after hatching, the brood is often taken to flushed areas that are rich in insects. These invertebrates represent a valuable source of protein for small chicks, and broods may travel up to 400 m daily to reach such areas (Hudson 1986a). Most chick mortality occurs in the first ten days after hatching, before full control of body temperature is attained. Chick survival varies between years and between areas. Possible explanations for this include the condition of the hen prior to laying (Moss *et al.* 1981), weather conditions after hatching (Erikstad 1985), and the availability of arthropod prey for the young chicks (Hudson 1986a). Grouse family groups break up as territories become re-established during the early autumn.

Usually, grouse are still in family groups on 12 August when shooting begins. The birds are driven by beaters so that they fly over the guns. Any disruption or disturbance at this point can reduce the number of birds that are shot (the bag). This is important because the estate's income is related directly to the bag. Grouse numbers are estimated each July and driven grouse shooting will only take place if the density is above about 60 birds km^{-2} (Hudson 1992). Occasionally exceptionally high densities are recorded; numbers in excess of 600 birds km^{-2} have been counted on moors in northern England (Hudson 1986a).

Cycling in numbers is a common feature of red grouse populations. The average cycle is four to five years in northern England and six to seven years in northeast Scotland (Potts *et al.* 1984). Extensive research has led to two main explanations. It has been suggested that cycles on English moors, with high rainfall, could be caused by the impact of the parasitic nematode *Trichostrongylus tenuis* on grouse breeding success (Hudson *et al.* 1985, Hudson and Dobson 1990, Hudson 1992). In addition, the sheep tick is a vector for the louping ill virus. Both diseases greatly increase grouse mortality. Elsewhere, on drier Scottish moors, it was demonstrated that changes in the social structure and territorial behaviour, within the grouse population, could cause cycles (Moss and Watson 1985, Mountford *et al.* 1990, Moss *et al.* 1996).

The most important predators of red grouse are foxes and crows. Although these are controlled vigorously by gamekeepers surprisingly little is known of their actual impact on grouse populations. Crows, foxes, stoats and weasels, can be legally controlled through snaring, trapping or shooting, and records of numbers killed suggest that populations of crows and foxes in Scotland may have increased in recent decades (Hudson 1992). In addition, several species of raptor prey upon red grouse. Grouse can form a major component of the diets of specially protected species such as hen harriers, peregrines and golden eagles. This has resulted in some conflict between the owners of grouse moors and conservationists. Grouse are taken less often by the common buzzard and goshawk, and only very rarely by short-eared owls, sparrowhawks, kestrels and merlins.

Throughout the twentieth century the red grouse has been in decline. Most of this decline can be explained by a decrease in the extent of heather moorland. In addition there is some evidence that grouse density has decreased on the remaining moors. Loss of grouse moor can be attributed primarily to three things: upland conifer afforestation, reclamation of heather to grass for more intensive agriculture, and poor heather management through excessive grazing and inappropriate burning. Inevitably a reduction in the amount of habitat leads to a reduction in total numbers of grouse although explanations for the reductions in grouse densities appear complex. The decline in density has varied between regions. In Scotland there has been a general reduction in numbers shot since before the last war. These reductions have occurred even on those estates that continue to manage for grouse. However, not all moors show this general decline. Falling standards of moorland management may be involved (Phillips and Watson 1995). Many grouse moors are grazed by sheep; others support cattle or red deer, and some a combination of all three. Heavy grazing reduces heather cover, standing crop, shoot production and perhaps most important, the heather's carbohydrate reserves. Sustained heavy grazing eliminates heather plants and rootstocks, and prevents seed production until the habitat is eventually lost (Lawton 1990).

Fire is an essential tool in moorland management because it produces a flush of tender, nutritious foliage. If burning is conducted properly the new vegetation is beneficial for both grouse and other grazing animals. Well-managed moors maintain heather by rotational burning to produce a fine-grained patchwork of short heather for grouse food and longer heather for cover. However, too frequent burning or badly controlled, large fires can lead to invasion by other plants. Burning standards have declined during recent decades, not least because there are now fewer shepherds and gamekeepers. Finally, there are theoretical reasons to believe that the fragmentation of heather moorland may lead to reduced grouse densities on the remaining, more isolated, patches of habitat. Isolated moors may be more difficult to protect from predators and grouse from adjacent moors may encounter problems in emigrating to them successfully.

One of the most topical issues in upland land management concerns the relationship between birds of prey and red grouse. Birds of prey are of great conservation interest, while grouse are important to the economy of many upland estates. The persecution of birds of prey, which stems from concerns about their impact on viable moorland management, has generated conflict between conservation and grouse shooting interests. In order to address these issues a five-year study, centred at Langholm in the Scottish Borders, was undertaken to provide an assessment of the impact that raptors have on the numbers and bags of grouse (Redpath and Thirgood 1997). The area had been managed as a high-quality grouse moor since the last century, and until the start of the study the annual bags had tended to fluctuate, with peaks approximately every six years. Additional, comparative, studies were made on five other Scottish moors. On most of these moors, as on many others, grouse bags had been declining for several decades. Throughout the study there was rigorous protection of Langholm's raptors. Other legally controllable predators (notably foxes and crows) were killed as before, and the heather was managed as usual. The research focused on the hen harrier and peregrine falcon, as these two species are of greatest concern to most grouse managers.

Recent surveys have aimed to find the total breeding numbers of hen harriers and peregrines in Britain (Bibby and Etheridge 1993, Crick and Ratcliffe 1995). The British harrier breeding population was estimated at around 630 females in 1988–89 and the total peregrine breeding population at around 1280 pairs in 1991. Grouse moors provide an important habitat for both species, particularly the harrier since most nest on heather moorland. Peregrine numbers have now recovered from the impact of organochlorine pesticides in the 1950s and 1960s to densities that are the highest in the world. In contrast, harrier numbers appear to be stable or declining, despite large areas of suitable but unoccupied habitat in certain parts of their range (Etheridge *et al.* 1997, Potts 1998). Harrier numbers were dramatically reduced by persecution before the end of the last century, and it is only since the Second World War that they have become re-established in mainland Britain. The recovery of the harrier population has coincided with the general decline in keepering levels and with the extensive afforestation programmes that provide suitable habitat in young plantations. Etheridge *et al.* (1997) estimated that, because of illegal control measures, both annual survival and breeding success of harriers were significantly lower on grouse moors than elsewhere. Such persecution may account for the fact that the national population has stopped increasing.

In early spring prospective harrier breeding sites are marked by sky dancing displays. Nests are built on the ground in tall heather and eggs are laid from mid April onwards (Redpath *et al.* 1997). Although harriers defend a small area around their nest site, separate pairs can nest within 500 m of one another. In some areas they are polygynous. Harrier chicks usually hatch in late May and start flying around mid July. Three weeks after fledging, the family groups leave their natal site. Harriers forage in a low systematic fashion taking their prey on the ground.

Peregrines tend to be highly territorial and pairs generally do not tolerate other peregrines breeding nearby (Ratcliffe 1993). They usually nest on cliff ledges, though they will occasionally nest on sloping ground, and the nests of different pairs tend to be evenly spaced throughout suitable habitat. They start breeding earlier than harriers, and usually lay three or four eggs in early April. Chicks hatch about four weeks later in early May and fledge after a further six weeks. Families stay together for at least two months after fledging, and then the young disperse. They usually catch their prey in the air.

The most rigorous experimental approach to the raptor problem would have used a paired, crossover design that compared the grouse numbers in the Langholm area, where raptors were protected, with grouse numbers in similar areas from which raptors were removed. The treatments would then be reversed after a period of years. This procedure was considered impractical and ethically unacceptable. Instead a less-powerful, non-experimental approach was adopted. This involved measuring everything necessary to assess the impact of raptors. The study did have the benefit of bag records from two other moors in the same region, where raptors were much scarcer than Langholm, but where the annual bags had previously fluctuated in parallel with those at Langholm.

Following protection from suspected illegal killing and other interference, the average density of breeding harriers increased for four years on four of the moors. For example, during 1992–96, harrier numbers at Langholm increased from two to 14 breeding females. Peregrine numbers were generally more constant over time, although at Langholm numbers increased from three to five or six pairs. During winter the numbers of peregrines and harriers varied considerably between geographical areas. At Langholm, a similar number of peregrine sightings were recorded each winter, but sightings of female harriers fluctuated in line with grouse density.

Harriers and peregrines appeared to hold the grouse population at a continuing low level, preventing the population from cycling but keeping post-breeding numbers too low to support driven grouse shooting. Langholm has a naturally high harrier/grouse ratio because sheep grazing has produced a mixture of heather and grassy areas that favours high densities of harriers. This situation may not hold on more heather-dominated moors elsewhere, which are less favourable to harriers. However, the Langholm habitat is typical of many heather/grass moors south of the highlands that have been affected by the same excessive sheep grazing in recent decades.

On the other five moors raptors were protected but gamekeepers controlled the numbers of foxes and crows. Each year, during 1992–96, the abundance of grouse, songbirds (mainly meadow pipits) and small mammals (mainly field voles) was estimated for each moor. The numbers, breeding success and diet of hen harriers and peregrines were also monitored. At Langholm grouse mortality and raptor hunting behaviour was also recorded, in addition to measuring a number of habitat features. Finally, records of grouse bags were examined to see how the number of grouse shot changed in the presence of breeding raptors.

Analysis of aerial photographs, showed that 48 per cent of heather-dominant vegetation was lost from Langholm moor between 1948 and 1988, mostly at lower altitudes. This loss of heather, and consequent increase in grass, was attributed to heavy grazing by sheep. Grouse bags on the same moor have shown a consistent and significant downward trend since 1913, and have also shown six-year fluctuations with the last peak in 1990. Given that raptor breeding densities were thought to be very low before 1990, it is extremely unlikely that raptors were responsible for either the long-term decline or the fluctuations in grouse bags.

On average spring raptor predation removed 30 per cent of the potential breeding stock of grouse, and in the summers of 1995 and 1996 harrier predation removed on average 37 per cent of grouse chicks. Most of these adult and chick losses were probably additional to other forms of mortality, and together reduced the post-breeding numbers of grouse by an estimated 50 per cent within a single breeding season. In each year, raptors also killed on average 30 per cent of the grouse between October and March, but it was not possible to determine what proportion of these grouse would have survived the winter in the absence of raptors. A simple, mathematical model of the grouse population at Langholm was developed. The model combined the estimated reduction in breeding productivity with the observed density-dependence in winter loss. In the absence of breeding raptors the model predicted that, over two years, grouse breeding numbers would have increased by 1.3 times and post-breeding numbers would have increased by 2.5 times.

Over the course of the study, there was no evidence that predation on adult grouse at Langholm was directly influenced by any of the habitat features measured. However, a greater than expected proportion of harrier attacks on grouse broods occurred in areas that were a mixture of heather and grass, as opposed to pure heather or pure grass stands.

Throughout the study, grouse density on Langholm moors in July averaged 33 birds 0.5 km^{-2} and numbers did not change significantly from year to year. Grouse bags did not peak in 1996 as expected from past records. In contrast, grouse bags on two other nearby moors, which had previously fluctuated in synchrony with those at Langholm, increased to high levels in 1996. These moors held only low densities of raptors. Predation by much larger numbers of raptors at Langholm was considered the most likely explanation for the continued low grouse density and low grouse bags on this moor during the study period. Bags on other moors where raptors were protected did not exhibit the same patterns as observed at Langholm. This was either because raptor numbers remained at low density, or because driven shooting was already not viable by the time raptor protection occurred.

Where raptors were not persecuted, breeding densities of harriers and peregrines varied considerably between different moors and were not related primarily to grouse densities. The highest breeding densities of harriers occurred on moors with a high ratio of grass to heather where alternative harrier prey such as meadow pipits and small mammals were most abundant. Peregrine breeding densities are lower in the Highlands than in the north of England, probably because of differences in the abundance of pigeons, which are their main prey. It was predicted that in the absence of persecution raptor numbers would be greatest on southern rather than northern moors and on moors with a high ratio of grass/heather. Extrapolating from data on harrier and peregrine diet, it was suggested that the impact of raptor predation within moors would be greatest when grouse densities go below approximately 12 pairs per km^{-2}.

The economic effect of predatory birds is apparent from some of the results of the Langholm study. Grouse numbers were reduced by up to 50 per cent, and the effect was most marked in 1996 when the expected peak in grouse numbers did not happen. It was expected that about 2000 grouse would be shot, the actual number was 67. It is estimated that this cost the estate almost £100000 in lost income. However, to place this loss in context Yalden (1981) estimated that the amount of grouse moor, in the Kinder–Bleaklow region of the Peak District National Park, had decreased from 154 km^2 before the First World War to 99 km^2 in 1979. The main cause of this reduction was thought to be land management changes, primarily more sheep grazing. Based on an estimate of 100 pairs of grouse km^{-2}, and an average of four young per pair, this represents an annual loss of over 30000 grouse.

One consequence of the Langholm study is that there is now a general consensus on the impact of predatory birds on grouse moors. Both conservationists and estate owners now accept that illegal persecution has suppressed the number of birds such as the hen harrier. It is also accepted that predatory birds can have a significant impact on estate incomes because the number of grouse available for shooting is reduced. The difficulty is finding a solution that is acceptable to all sides. The predatory birds are protected by European Union legislation (the Birds Directive) so it would be very difficult to control their numbers by legal culling. The long-term solution is probably related to careful habitat management. In the short term more *ad hoc* measures, such as the provision of extra food for breeding raptors, may be necessary.

FORESTRY

Britain's forests have experienced hundreds, if not thousands, of years of unsustainable exploitation. The problem had been recognized during the seventeenth and eighteenth centuries when new plantations were established. Unfortunately this new resource was insufficient to cope with the timber demands of the industrial revolution during the nineteenth century. The poor state of Britain's forests became very apparent during the First World War when woodland covered only 5 per cent of the land area. Inevitably Britain experienced a shortage of timber, an important strategic resource.

As a direct consequence of these timber

shortages the Forestry Commission was established in 1919 with the aim of preventing similar problems. The first of its two main functions was to advise and implement Government policy for all forests and to promote the interests of forestry. The second function was to manage almost half of Britain's forests. In this management capacity it was responsible for producing timber, conserving the forest environment and providing recreation opportunities. In 1992 the Forestry Commission was re-organized to clarify the distinction between the two functions. Its regulatory, research and grant-aiding functions are now undertaken by the 'Forestry Authority' whilst the 'Forest Enterprise' has responsibility for the management of the forests.

Even after the timber supply problem had been recognized natural forests continued to be cleared, partly because of the inevitable delay between planting and the first harvest date. Because the main focus of the Forestry Commission was timber supply little thought was given initially to conservation. For a variety of ecological, agricultural and economic reasons the best place to plant these new forests were the upland margins and the best species to plant were mainly introduced conifers such as the sitka spruce from the west coast of North America. After the Second World War government policy prevented forest planting on inbye land, consequently new plantations were pushed further into the uplands (Zehetmayr 1987). Recently there has been some change of emphasis (Forestry Authority 1998). In particular there is a recognition that new woodlands need to move 'down the hill', partly to take land out of agricultural production so that food surpluses can be reduced.

Since the establishment of the Forestry Commission woodland cover has doubled to about 10 per cent, a figure that is still low in comparison with most European countries. Little of this woodland is natural and almost 70 per cent is coniferous plantation. Within the United Kingdom woodland cover is distributed unevenly, for example over half is within Scotland (Table 3.5). Tudor and MacKey (1995) calculated that, between 1940 and 1970, coniferous forest increased by over 4500 km^2 in Scotland. This was largely at the expense of heather moorland (2741 km^2), unimproved grassland (1194 km^2) and blanket mire (1064 km^2).

The establishment of a forest produces a sequence of environmental changes. The most obvious effect is the loss of open country. There are other changes arising from:

- the removal of grazing and cessation of burning;
- increased drainage;
- the application of fertilizers.

Table 3.5: Area (1000 ha) of coniferous plantation in 1997

Country	*Forestry Commission*	*Private woodland*	*Total area*
England	169	214	383
Wales	110	58	168
Scotland	469	512	981
Northern Ireland	56	20*	76*

Note: *The NI data does not differentiate between conifer and broadleaved forests

Source: Forestry Commission data

Initially the ground flora becomes taller and there may be some changes in the species composition because the soil is drier and the removal of grazing and burning alters the competitive balance between the plants. After about 15 years the trees begin to mature and the canopy closes to produce a very dense cover that restricts the ground flora by extreme shading. Trees are usually thinned to produce a crop of tall trees with straight trunks. If soils are thin there may be no thinning because this could increase the risk of windthrow. Mature trees are felled at between 35 and 80 years old, depending on the amount of thinning (Mason and Quine 1995). Because the forest is planted in even-aged blocks clear-felling is used to produce open spaces. These are quickly replanted. The ground vegetation during the early phases of the second rotation crops passes through a series of stages that may differ from natural plant assemblages. Eventually the canopy closes again and future development of the ground flora is halted (Wallace and Good 1995). Similar changes occur with the invertebrate communities. Butterfield *et al.* (1995) demonstrated that although conifer plantations can contain as many carabid species as the surrounding open countryside most of these are found in the clear-felled areas. Some of the rarer species, particularly those associated with wetter ground, tend to be missing from the plantations.

During the 1970s some people began to get concerned about the effects that the widespread afforestation of upland regions was having on the ecology of the uplands. In some regions, such as Argyll, forest covers more than 50 per cent of the uplands (up to 600 m). The most obvious losers were the moorland birds and the acidification of upland streams (Table 3.6). Part of the problem lay in the tax incentives for forest planting. It was very profitable to plant new forests. The detrimental effects of inappropriate forest schemes were very apparent in the Flow Country of Caithness and Sutherland. The Flow Country is a region of peatland that has national and international conservation value. By the mid 1980s media campaigns against these forests began to affect government policy. Even so, the Secretary of State for Scotland approved SSSI

Plate 3.5: Conifer afforestation – unsympathetic planting of peatland in Connemara, western Ireland

Table 3.6: The ecological effects of afforestation in the uplands

Effect	*Notes*
Water acidification	There is very strong evidence that plantation forests alter the chemistry of streams that flow through them. For example, Omerod *et al.* (1993) found higher aluminium concentrations in streams draining conifer catchments compared with those draining open moorland or deciduous woodland catchments. Water acidity is also known to rise because conifer forests exacerbate the effects of acid deposition (Wright *et al.* 1994). Increasing acidity and aluminium concentrations are known to decrease the number of trout and the taxonomic diversity of stream invertebrates (Omerod *et al.* 1993, Rees and Ribbens 1995). The changes to the stream invertebrates are also thought to be responsible for the decline in some water birds such as the dipper. The acidification of freshwater on at least 40 SSSIs has been linked to afforestation in the catchment (Rimes *et al.* 1994).
Loss of open habitat	Some of the most obvious effects of afforestation relate to the displacement of the internationally important upland bird community that is associated with open moorland. These effects were reviewed by Thompson *et al.* (1988).
	Loss of breeding opportunities for ground nesting species restricts birds such as waders, grouse and some merlin.
	The loss of open ground restricts many predatory and scavenging birds because both prey and carrion are less abundant. Species such as golden eagle, raven, merlin, hen harrier and short-eared owls are particularly at risk. However, it is also known that both hen harriers and short-eared owls are able to exploit the pre-thicket stage of plantations when vole numbers usually rise as a consequence of ground vegetation changes. Unfortunately the birds do not make similar use of replanted parts of the forest. It appears that cleared patches will only be used if they exceed 62 ha (Shaw 1995).
Changes to the surrounding habitat	The effects of a new forest extend beyond its perimeter fence. There will be fewer intentional fires close to the forest and grazing pressure often decreases. In addition the forest drainage system may begin to dry out the surrounding land. For example, Lavers and Haines-Young (1997) thought that edge effects from new plantations in the Flow Country were a serious threat to the remaining dunlin. Less obviously, the forest may create refuges for predators that exploit prey in the surrounding open countryside. These wider effects are difficult to quantify since they depend on many factors such as the size and shape of forest patches and how they are positioned with respect to the wider landscape.
Benefits	The forest canopy creates additional habitat for forest birds. However, few of these are threatened species. The birds that are displaced by the forest are at greater risk.
	Although conifer plantations are generally thought to be detrimental to birds of open moorland there is some evidence that the merlin has adapted to using old crow nests close to the edge of the plantation (Parr 1994).

designation for only 50 per cent of the remaining unforested peatland. The most important change came about in the March Budget of 1988 when alterations to taxation and forest grant schemes made forestry a less attractive investment. The designation of the Flow

Country peatland SSSIs as a SAC should ensure that they receive even greater protection in the future. Recently there has been a move towards planting more native species, often including broad-leaved trees. There has also been a recognition that wildlife benefits from a greater structural diversity within the forest.

ATMOSPHERIC POLLUTION

Anthropogenic changes to the composition of the atmosphere have important ecological effects that are often magnified in upland environments. Most of these changes to the atmosphere are linked to industrialization and lead to an increase in the concentrations of some natural and some novel atmospheric components. The term air pollution is generally applied to chemicals such as sulphur dioxide that give rise to 'acid deposition' which can change the chemistry of soils and freshwater. The increases in atmospheric carbon dioxide and other 'greenhouse' gases are usually considered separately because their effects are thought to be climatic.

There are several well-known ecological consequences of air pollution in lowland and urban areas, for example industrial melanism in the peppered moth *Biston betularia* and the two-spot ladybird *Adalia bipunctata*. It has been recognized for a long time that uplands close to industrial areas, such as the South Pennines, have also experienced significant ecological changes as a consequence of air pollution. One of the most depressing findings of recent studies is that organisms living on the mountains of the NW Highlands of Scotland are not free from the effects of air pollution. Upland environments are particularly sensitive because they are frequently covered by clouds and mists that enhance the transfer of pollutants from the atmosphere to the plants. From the mid-nineteenth century up to the late 1950s the major pollutants were smoke and sulphur dioxide. These had independent effects, for example the smoke particles blackened surfaces whilst the sulphur dioxide killed some lichens and bryophytes. It seems likely that air pollution was the main factor leading to the loss of *Sphagnum* from mires on the hills between Lancashire and Yorkshire (Lee *et al.* 1988). It is also possible that particulate

Plate 3.6: Eroding blanket bog – south Pennines

pollution contributed to the conversion of heather moor to purple moor grass dominated grassland in regions such as Wales and southern Scotland (Chambers *et al.* 1979). The impact of severe air pollution on people living in cities was responsible for the introduction of the Clean Air Act, which resulted in a decrease, but not a complete halt, to this type of pollution.

More recently the role of nitrogenous components in acid deposition has become increasingly apparent. It is possible to follow the historical build-up of nitrogen by examining plant material stored as herbarium specimens in museums. The evidence suggests that the *Racomitrium* heath community, which is one of Britain's largest natural communities, is under threat from air pollution. Baddeley *et al.* (1994) found that the nitrogen content of *R. lanuginosum* collected from the summit of Ingleborough in Yorkshire increased from 4.6 mg g^{-1} in 1879 to 12.3 mg g^{-1} in 1989. *R. lanuginosum*, especially when growing as extensive mats on mountain summits, receives most of its nutrients from the atmosphere (ombrotrophic). Consequently, an increase in the amount of atmospheric nitrogen may have direct effects on this and similar plants. The Ingleborough results, and results from transplantation studies, have shown that enhanced atmospheric nitrogen is reflected in a higher nitrogen content in *Racomitrium* which Baddeley *et al.* (1994) think may have a number of effects. The excess of supply over demand could alter the balance of nitrogen supply between plants, resulting in a change to the competitive balance. There is some evidence that this, in combination with increased sheep grazing, is the explanation for the loss of *Racomitrium* heath in areas south of Glasgow. The extra nitrogen could also accelerate the microbial decay of dead plant material, a process that would also favour grasses.

Racomitrium heath is not the only threatened montane bryophyte community. Woolgrove and Woodin (1996) studied the rare and fragile bryophyte communities of late snowbeds which, nationally, cover about only 250 ha. Because of the particular characteristics of late snowbeds these communities experience more air pollution than other montane species. The problem is that snow scavenges pollutants from the air and then releases these at enhanced concentrations during brief 'acid flushes' that damage endangered species such as *Kiaeria starkei*.

Streams and lakes in the uplands are also affected by air pollution, the main effect being water acidification, which has an effect on the species composition of upland streams (Omerod *et al.* 1993). As with the plants, the pollution is not restricted to the vicinity of industrialized areas. Battarbee *et al.* (1996) examined the sediments in Lochnagar, a small but deep loch close to the summit of Lochnagar on the southeastern fringe of the Cairngorms. Using sediment cores they were able to show that the diatom flora had changed over the last 100 years, suggesting an increase in acidity from pH 5.6 to 5.0. The gross effects of air pollution were apparent in a sediment colour change from light grey to very dark, a consequence of contamination by high concentrations of fly-ash from fossil fuels. Changes in emission regulations may help to reduce this type of contamination but it is now apparent that atmospheric nitrogen deposition is one of the major causes of water acidification in the uplands.

Because temperature has such an overriding influence on the upland habitat, global warming is potentially one of the greatest threats to the upland environment. The causes of global warming, such as the burning of fossil fuels and the release of methane by agriculture, are reasonably well understood. Even if large changes are made to greenhouse gas emissions it seems likely that climate change will continue. Any changes to the average temperature will have rapid, and possibly unmanageable, consequences for the ecology of the uplands. The consensus opinion from the Intergovernmental Panel on Climatic Change is that Britain's average temperature will increase by about 1.3°C by the year 2050. This is equivalent to a southern latitudinal shift of about 300 km or a decrease in altitude of around 200 m. Consequently global warming will move upland ecological boundaries to higher altitudes. It is quite possible that montane environments will 'fall off' the mountaintops. Once this has happened the process cannot be reversed since there will be no refuges from which re-colonization can take place.

Even if the climatic predictions from the various models are correct it is far less certain how the flora and fauna will respond to the changes. As with most ecological processes the reality will not be as simple as first appears. For example, elevated temperatures will have disproportionate effects in the more oceanic regions of Britain because oceanic regions have relatively flat annual temperature curves. Consequently a small rise in temperature will have a larger effect on the length of the growing season than would a similar temperature rise in a more easterly continental location (Pepin 1995).

In the uplands the most important climatic effects are likely to be an increase in extreme events, such as: high winds and unseasonable frosts; changes in evapotranspiration leading to less soil moisture, and a marked reduction in snow cover. The effects of such changes in the British uplands are likely to be great because of the unique mixture of species, some being at their northern limits, whilst others are at their southern limits. There will be both winners and losers in this process that will be reflected in the establishment of new communities.

Although it is possible to suggest some broad consequences of global warming for upland communities and species, local circumstances are likely to result in considerable variation. The area of land currently described as sub-montane and montane habitats will be decreased. Because of species–area relationships a simple reduction in the area of upland habitats should lead to the eventual loss of some species and a decrease in upland biodiversity, even if new species colonize the warmer uplands. The structure of blanket bogs could alter as they become drier and warmer. Some montane plant communities may be lost, snowbed communities will be at particular risk. Birds such as the dunlin and curlew, which move off the uplands to overwinter around the coasts, may suffer because of changes to, or destruction of, their winter habitat. A rising sea-level, combined with the retention of current coastal agricultural land, may mean that very little winter habitat survives. Similar problems may be expected for species that migrate to other countries during the winter. Some specialist montane birds and mammals such as the dotterel, snow bunting, ptarmigan and mountain hare may become locally extinct.

However, there is a different scenario for the British climate that could lead to an expansion in upland habitats. There is no doubt that, in comparison with other locations at a similar latitude, Britain has mild winters. This is almost entirely a consequence of the transport of warm water to our western coasts by the North Atlantic Drift. This current is driven by a complementary southern movement of cold arctic water. It is unclear how the North Atlantic Drift will respond to a changing global climate. If it moves further south, towards France and Spain, our winters may become colder even though global temperatures are rising. Whichever climatic scenario applies to the United Kingdom we can be certain that there will be changes in the amount of upland habitat and its species composition. We should have the answers in the next 50 years.

RECREATION, ACCESS AND SKIING

The last 20 years or so have witnessed ever increasing recreational use of the countryside, much of it concentrated in highly popular upland locations, e.g. the Peak District, the Lake District and the Cairngorms. The most important activities include rock climbing, hill walking, mountain-biking, skiing and angling.

On many of the most popular mountains and moorlands hill walking and cycling are responsible for the growing problem of path erosion. On wet peaty ground, e.g. the Pennines, the damage is particularly serious since repeated use and extreme winter weather prevent the vegetation from recovering. As the footpath surface deteriorates walkers tend to follow a new parallel route. This considerably widens the original path. Recently concerns have been raised over the possible effects of disturbance to moorland breeding birds (Sidaway 1990). This subject has generated a great deal of controversy especially in the context of the Peak District National Park. Sound ecological data are however rather limited and equivocal. Attempts to quantify the adverse effects on breeding golden plover have been described by Yalden and Yalden (1989, 1990). The problems appear to be localized and confined to a relatively few highly popular locations. However, there is increasing concern that parties of birdwatchers may be creating some disturbance to upland birds. Staines and Scott (1994) reviewed the potential disturbance of red deer by recreational activities, and the implications for deer management. Although they found no evidence that populations of red deer were adversely affected by increased disturbance some estates have reported that the 'quality' of stalking was diminished by regular disturbance and that culls were less selective. The problem is that even a few walkers, particularly if they have a dog, can cause the deer to disperse making culling and stalking more difficult.

Access to moorlands and heatherlands is a particularly topical issue. Whilst wild life conservation is clearly of great importance there is always the danger that unscrupulous landowners will use spurious conservation arguments to exclude walkers from their land. There is a strongly held view by many of the public that favours a presumption of public access on foot to all open country as of right, with local restrictions only where clearly justified.

Initially rock climbing was concentrated on gullies with vegetation being removed to facilitate better holds. This sometimes destroyed colonies of rare alpine plants. More recently climbers have tended to utilize 'hard rock' with little vegetation. This has increased disturbance at some seabird colonies and sites containing breeding peregrines, ravens and golden eagles. Some breeding birds have adapted to climbing disturbance, whilst others have moved to quieter alternative cliffs. In Scotland the climbing resource is much greater and competition for crags between birds and climbers is much more localized. Problems occur more frequently in England and Wales. However for many popular routes with nesting sites of vulnerable species seasonal restrictions have been negotiated between statutory conservation agencies and the access and conservation committee of the British Mountaineering Council. These usually cover the period 1 April – 31 July but may be amended locally

Plate 3.7: Paved footpath – an extreme solution to some of the problems of too many visitors (Cairngorm)

in the light of detailed information regarding particular breeding birds (Sue King, *pers comm*).

The Scottish ski resorts differ from those in most other countries because most of the skiing takes place above the tree line in the fragile montane habitat. Adverse ecological impacts from skiing may result from skiing itself and from the many ancillary works associated with ski field development. Some of the earlier ski developments in Scotland were insufficiently site-sensitive with tracks and pistes being bulldozed. This produced considerable habitat modification and some local erosion. More recent developments have made use of helicopters to reduce vegetation damage. Skiing, particularly towards the end of the season on bare ground, can damage vegetation. This is detrimental to both the vegetation and the ski resort since a good ground cover is important for snow retention. One of the most important concerns is year-round use of ski tows and chair lifts which makes access to formerly remote terrain, such as Cairngorm plateau, much easier. If the current trend for less winter snow continues, site operators will be forced to further diversify their business and encourage more summer use of the high tops. Bird and animal numbers near ski lifts have been examined by Watson (1979, 1982, 1996b) whilst soil erosion and damage to vegetation have been assessed by Bayfield (1974) and Watson (1985). Some decline in ptarmigan numbers in a part of the Cairngorms has occurred as a result of collisions with ski-tow cables. More crows and gulls have been attracted to the higher ground by discarded food but their predation impact on important breeding bird populations is as yet limited. After much controversy, detailed monitoring programmes on an annual basis have been suggested for most of the ski centres (Bayfield *et al.* 1988).

Anglers use most upland lakes, rivers and streams at some time during the spring and summer. Disturbance to at least 20 species of breeding bird results from fishing but little data on adverse impacts are available (Ratcliffe 1990a). Anglers, keeping birds off their nests, increase the predation risks to eggs and may in part be responsible for poor breeding success of both black-throated and red-throated divers in some areas. Local declines of the common sandpiper have been noted (Holland *et al.* 1982), whilst breeding oystercatchers, ringed plover, dipper and grey

Table 3.7 A summary of the major human impacts on upland habitats

Impact	*Effects*
Excessive grazing	Restricted forest regeneration and spread of bracken
	Loss of dwarf shrub habitat, possibly reducing grouse bags
	Loss of *Racomitrium* heath from some mountain summits
	Plants that are sensitive to grazing are restricted in inaccessible locations such as cliff ledges
Agricultural practices	Burning
	Changes to the species composition on dwarf shrub heath
	Damage to blanket bog
	Draining
	Loss of wetlands including bogs
	Increased soil erosion
	'Reclamation'
	Conversion of semi-natural habitat to semi-improved and improved grassland
Afforestation	Loss of semi-natural habitats
	Reduction in birds of open moorland
	Changes in water chemistry
Pollution	Loss of sensitive plants, especially montane bryophyes
	Increased soil and peat erosion
Recreation	Localized damage to vegetation
	Localized increased soil erosion
	Some disturbance of wildlife

Source: Based on Thompson and Horsfield 1997: Table 3

wagtail may all be susceptible to increasing disturbance (Ratcliffe 1990a).

Recreational impacts on the natural heritage have been reviewed by Sidaway (1994), who concluded that, to date, most studies concentrated on short-term rather than the more crucial long-term effects. A great deal more research is clearly required if the impacts of recreational disturbance on animal populations are ever to be fully teased out. In the short term education, tolerance and a willingness for all interested groups to actively cooperate at the local level is essential. Ultimately, the best way to reduce recreational damage probably depends on the management of people rather than focusing on habitat impacts.

SUMMARY

The impacts described in this chapter are summarized in Table 3.7

4

CASE STUDIES

•

The case studies have been selected to illustrate, in some detail, the important processes and factors that have been identified in the previous two chapters.

RED DEER

Following the demise and extinction of the elk 9500 years ago the red deer *Cervus elaphus* became the largest wild mammal in Britain. Although it was once abundant in the formerly extensive native woodlands its range began to contract with forest clearances that were associated with human settlement from about 5000 BP. Agricultural intensification, combined with the widespread introduction of sheep and hunting, further reduced the numbers and range of red deer. The low point was probably reached in the mid to late eighteenth century (Cameron 1923). Most British red deer are now found in Scotland, although there are more isolated populations in regions such as the Lake District and Exmoor. There are also a number of captive groups in country parks and estates such as Lyme Park on the outskirts of Stockport.

The number of red deer increased during the nineteenth century as sheep farming declined in profitability and deer stalking became a popular sport. Large tracts of land were acquired specifically for sporting purposes and by the early part of the twentieth century more than 1.5 million ha were managed mainly for deer hunting (Callander and Mackenzie 1991). These largely treeless 'deer forests' now contain the bulk of the red deer population, although increasing numbers are beginning to use conifer plantations. Since the early 1960s the Scottish population has increased dramatically, doubling to 300000 (Figure 4.1), possibly related to a combination of mild winters, dry summers and under-culling of hinds (Clutton-Brock and Albon 1989). Increases in

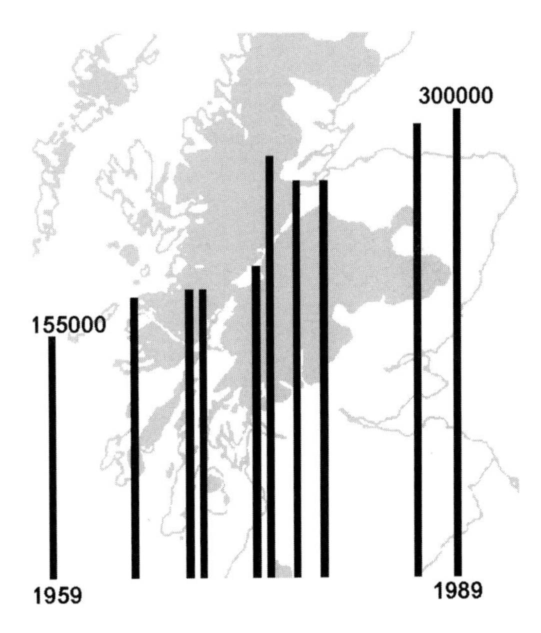

Figure 4.1: Recent changes in red deer numbers (based on RDC data). The background map illustrates the major red deer regions

Source: Based on Whitehead 1964

the deer density and an expansion of the range have taken place so that red deer now occupy more than 3 million ha or 40 per cent of Scotland (Scottish Natural Heritage 1994). The impact that this population of large mammals has on the ecology of the uplands is very significant.

Culling

Red deer are culled extensively in Scotland for sport and to protect agriculture and forestry interests. In the absence of predators, such as the wolf, culling has become a very important aspect of red deer ecology. The total numbers of red deer shot annually increased from about 24000 in 1973 to around 70000 by 1993 (Red Deer Commission (RDC) in Scottish Natural Heritage 1994). At this higher level the cull figures exceeded recruitment for the first time since the inception of the RDC. Within the Highlands between 10 per cent – 17 per cent of stags and 6 per cent – 12 per cent of hinds are killed each year (Clutton-Brock and Albon 1989). Stags and hinds are shot during separate stalking seasons that run from 1 July to 20 October and 21 October to 15 February respectively.

The extent to which the red deer population in Scotland is approaching or exceeding the upland carrying capacity is unclear. In the absence of high rates of culling the extent of possible further increases in population size remain unknown. However detailed studies (Mitchell *et al.* 1977, Clutton-Brock and Albon 1989) have indicated that population size can be limited naturally by changes in reproductive output. This is because there is a feedback mechanism that links an increase in population density to a decline in adult fertility and the survival of juveniles. This 'density dependence' has important implications for the sustainable management of red deer and other animal populations.

The treeless landscape, that is the typical British red deer habitat, is not their true habitat. Over time red deer have adapted to the change from woodland to open habitats by reducing body size, growth rates and reproductive output (Blaxter *et al.* 1974, Mitchell *et al.* 1977, Clutton-Brock and Albon 1989). In these open habitats deer tend to utilize similar vegetation to sheep, mainly grasses and dwarf shrubs, but their impact is far from clear. Mitchell *et al.* (1977) suggested that, in the absence of sheep, red deer grazing would bring about a reversion from grasslands to dwarf shrub heaths. Watson (1989), however, takes the view that high deer numbers are already resulting in a significant shift from dwarf shrub heath to grassland. It is certain that red deer make a substantial contribution to the maintenance of the open habitats that favour highly valued species such as the greenshank *Tringa nebularia* and golden plover. In addition deer corpses provide a valuable source of food for golden eagles, ravens and a variety of mammals (Scottish Natural Heritage 1994). Upland woodlands may benefit from some light grazing by deer. This is because their grazing can produce and maintain plant structural and species diversity (Mitchell and Kirby 1990). Niches for tree seedlings may also result from ground disturbance brought about by grazing animals. However, if there are too many deer excessive grazing will prevent woodland regeneration.

For over 150 years there have been conflicts between deer and agriculture. In some areas damage to crops is considerable but the overall economic impact of marauding deer on cropland is unknown. During the past twenty years about 3500 deer per annum (7 per cent of the total red deer cull) have been culled for agricultural protection.

Conflicts between forestry and red deer were initiated with the establishment of the Forestry Commission, although the full extent of the problem was not revealed until the 1950s. Red deer are now resident in many Scottish plantations (Staines and Ratcliffe 1987) but their numbers are difficult to census and estimates vary between 27000 and 50000 (Ratcliffe 1987). Damage to trees, by browsing and bark stripping, varies with factors such as tree species and size, the type of shoots browsed and the browsing frequency (Staines and Welch 1989). Although the overall economic impact of red deer damage to upland forests has not been fully quantified, it seems likely that relatively high culling rates of between 18–25 per cent are required to stabilize woodland deer populations (Ratcliffe 1987).

Native woodland, such as the internationally important Caledonian pinewoods, now covers less than 2 per cent of Scotland and in many of these woodlands little or no regeneration is currently taking place because of browsing (Mackenzie 1987). Studies by Watson (1983) indicated that there has been little regeneration for over 200 years in some of these woodlands and many of the trees are now approaching the end of their reproductive lives. Defining winter densities of deer, below which tree regeneration will occur, is difficult. Regeneration of Scots Pine has been observed with less than 5 deer per km^2 (Holloway 1967) and regeneration of birch and Scots Pine has been prevented at more 33 deer km^{-2}. Factors other than deer density also play a role in tree regeneration, and these include sheep grazing, ground cover, altitude, location of seed sources and soil conditions (Staines and Welch 1989). However, from the frequency of observed tree regeneration in enclosures, it is clear that grazing is an important factor preventing forest regeneration (Scottish Natural Heritage 1994).

Although red deer are an intrinsic part of the natural heritage, their rapidly increasing numbers are probably bringing about unwanted vegetation change through grazing and browsing. However, it is difficult to separate out the effects of deer from those caused by the 2 million sheep (Scottish Natural Heritage 1994) that share much of the same range in Scotland.

Scottish heather moorland declined in area by about 18 per cent over the period 1947–73 (Scottish Natural Heritage 1994). Over half (53 per cent) of this loss resulted from the conversion of heather moor to semi-improved (5 per cent) and unimproved grassland (48 per cent). Since heather tends to undergo succession to grassland under a combination of medium to high grazing with burning the change to grassland probably reflects the effects of excessive grazing. Because the range dynamics of red deer are complex (Welch *et al.* 1992) it is difficult to isolate the relative importance of sheep and/or burning when assessing vegetation change. Hill sheep are often intermingled with red deer at densities between <10 to >200 sheep km^{-2}. At lower altitudes it is known that 100–200 sheep km^{-2} may cause heather decline. At higher elevations a similar change can be brought about by less than 50 sheep km^{-2}. Because grazing is now so extensive, examples of near natural vegetation, which display altitudinal continuity from woodland through scrub and dwarf shrub heaths to montane summits, are extremely rare.

A proportion of many deer forests is burnt each year to remove dead vegetation and stimulate the growth of new shoots for deer, sheep or grouse. This activity is termed muirburn in Scotland and is ideally carried out on a rotational basis (Phillips *et al.* 1993). However the standard of muirburn is often poor. For example, fires may burn over large areas, or burn so hot that they damage the soil and roots, or burn at high altitudes or on steep slopes. All of these problems increase the potential for soil erosion and retard the regeneration of palatable vegetation. More importantly unpalatable species, particularly purple moor grass, cotton grass and deer sedge *Scirpus cespitosus*, may spread at the expense of those dwarf shrubs most utilized by red deer during winter.

Estates use deer-proof fencing for two purposes. First it keeps deer out of forestry plantations. Second it prevents them from straying onto adjoining land where they could cause some agricultural damage. Unfortunately fencing is a fairly crude and often unreliable tool for managing red deer populations, for example in deep snow fences are ineffective as deer simply walk over them. Confined deer may also suffer because they are prevented from making optimum use of their traditional range and movements between areas are restricted. Collisions with the fences also contributes significantly to the mortality of other species such as the black grouse and the rare and declining capercaillie (Baines and Summers 1997). Deer fencing has also raised other concerns including access restrictions, damage to sensitive habitats associated with fence erection and maintenance and the visually obtrusive nature of unnatural sharp boundary lines (Watson 1993a). Nevertheless, in the short term, fencing, coupled with culling, can assist natural regeneration of native woodland. Where land is managed primarily for deer stalking woodland regeneration is unlikely to occur in the absence of fencing.

Scottish Natural Heritage (1994) has made a number of recommendations for the future management and control of red deer populations.

- Reduce the red deer population by 100000 animals as a first step towards integrated management. Optimum populations should be defined at the local level and included within future management plans.
- Integrated approaches to deer management should be adopted throughout their range. The RDC and Deer Management Groups should be in the lead to achieve this, while Scottish Natural Heritage provides advice on natural heritage matters.
- Ensure that previous work on the interaction between grazing animals (mainly sheep and deer) and their effects on habitat are used to provide practical advice for improving management of the deer range.
- Identify woodlands of high priority, that as a result of deer grazing and browsing, are not regenerating.
- Culling and fencing are not to be seen as opposing alternatives, but as complementary management tools, to be used as circumstances require.
- Consider and investigate alternative methods of culling, such as live capture and contraception.

THE CAIRNGORMS

Many people consider that the Cairngorms are not only the most important mountain range in Britain, they are also one of the most important areas for conservation (Thompson *et al.* 1995b). Walking in the Cairngorms can be both an exhilarating and exasperating experience. The exhilaration derives from the landscape and the wild life (Nethersole-Thompson and Watson 1981). The exasperation is a result of the current patchwork of land ownership, linked with the varying levels of habitat protection and damage. Many international ecologists think that the Cairn-

Plate 5: A ski-development – Glenshee

Plate 6: Mountain hare – an important item in the diet of many golden eagles

Plate 7: Lough-studded blanket bog, western Ireland

Plate 8: Bracken invading heather moor, northern England

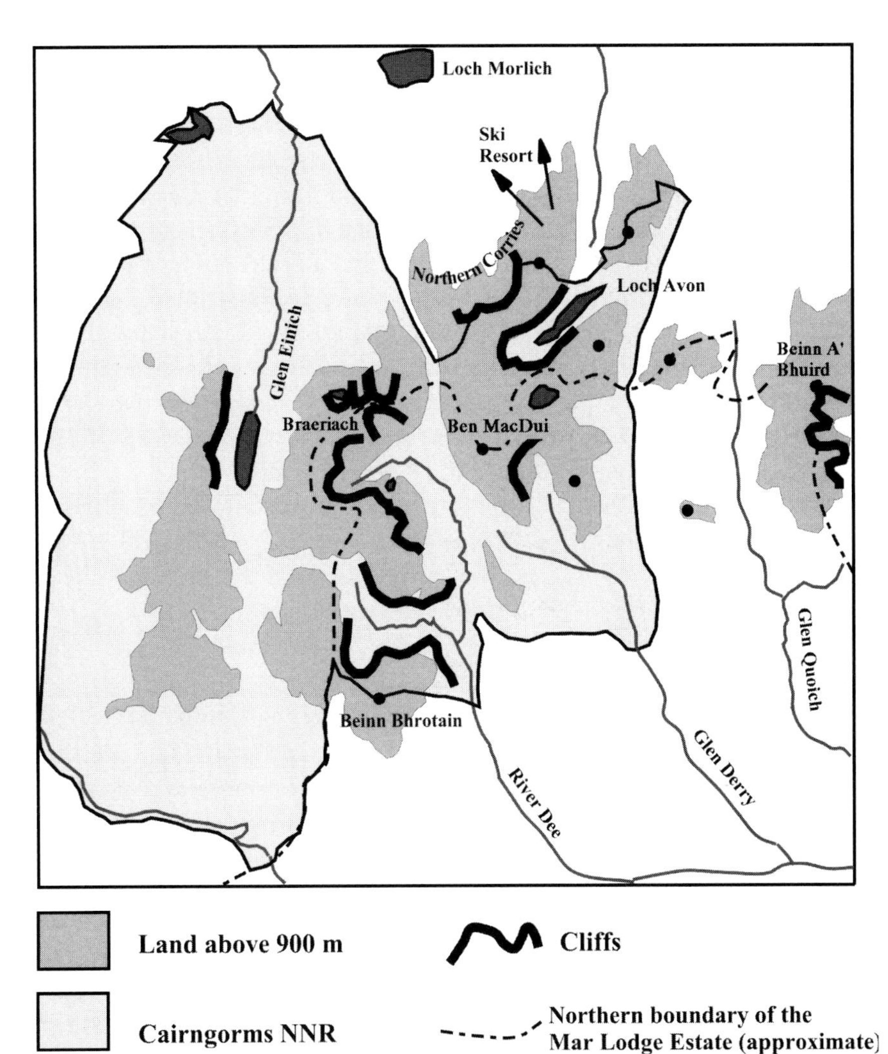

Figure 4.2: The Cairngorm mountains

gorms are of World Heritage Status, a designation that they would share with other globally significant habitats such as the Great Barrier Reef. Over 50 years ago the Ramsay Committee defined five areas, including the Cairngorms, for designation as Scottish National Parks. Unfortunately, and despite adequate funding, none were created although they did receive some protection from uncontrolled developments (Weir 1993). Recently the government has again raised the possibility of National Parks in Scotland but they have not yet put the Cairngorms forward for World Heritage Status designation, presumably because the bid would almost certainly fail given the current unsatisfactory levels of protection (Scott 1993).

The Cairngorms (Figure 4.2) are a roughly circular region between the Spey and Dee rivers, with a diameter of approximately 75 km (4050 km^2). The region includes four of Britain's five highest mountains (Ben MacDui, Braeriach, Cairn Toul and Cairngorm) and Britain's largest (260 km^2) National Nature Reserve (Taylor 1993). The NNR incorporates most of the high plateau around Ben MacDui and the surrounding summits. The geology is predominantly granite that weathers to form a coarse, nutrient-poor soil that makes the land unsuitable for domestic grazing. They are an unusual range of mountains because the summits are part of a relatively flat plateau that is dissected by deep valleys. The landscape is almost entirely a consequence of glaciation. Indeed, this is one of the most diverse sites in the world for glacial features. These geomorphological features are sufficient in themselves to make the region of international significance.

The Cairngorms are special because they have the best example of a continuous gradient of natural vegetation from deciduous woodland, through coniferous forest, to montane vegetation and fell field habitat. The plateau has a large expanse of montane habitat which, excluding Sweden, is the largest arctic-like block in the European Union (Watson 1996a). The montane flora includes Britain's best developed lichen-rich heath, a habitat more usually associated with European continental mountains (Thompson *et al.* 1993b). The montane fauna is also outstanding, with many species that are typical of arctic-alpine environments. Although the montane flora and fauna are sufficient to make the Cairngorms important, the valleys (glens) also contain important habitats. In particular there are some of the best remnants of the once extensive Caledonian pine forest including the forests of Abernethy, Rothiemurchus and Mar. There is now a possibility that improved management, in particular control of red deer grazing, will allow these isolated fragments to merge to form a very impressive forest that would be one of Britain's most significant conservation achievements.

The vegetation cover of the Cairngorm region seems to have changed little during the last 50 years (Hester *et al.* 1996). In a study of a 1000 km^2 region natural succession occurred in only 5 km^2. The largest changes were associated with the extension of forestry plantations, including the planting of 55 km^2 of moorland and the conversion of 11 km^2 of natural pine forest to plantation. When the Cairngorm ski resort was developed in the early 1960s there was an inevitable increase in the disturbance to what was previously a relatively quiet location. It seems that this disturb-

Plate 4.1: Montane plateau – an artic-alpine environment (Ben MacDui, Cairngorms)

ance affected the distribution of grazing animals so that grazing and browsing were reduced in the Northern Corries SSSI, between the ski resort and the Lairig Ghru. French *et al.* (1997) think that this relaxation in the grazing pressure allowed a natural subalpine scrub to develop, including some scattered pines growing at altitudes up to 830 m, almost 150 m above the supposed tree line. They suggest that one of the factors limiting further tree development is seed supply, which is related to the distance to the nearest forest.

Undoubtedly the climate of the Cairngorms is a significant factor in their importance. The following account is based almost entirely on McClatchley (1996) who integrated climate data from a number of disparate sources to get a better understanding of the weather in the Cairngorms. At the time of writing, 'live' data from an automated weather station on the summit of Cairngorm (1245 m) can be obtained via the Internet from http://www.phy.hw.ac.uk/resrev/aws/weather.htm.

The temperature lapse rates calculated for the Cairngorms are high at 7 to 10 °C 1000 m^{-1} during day time. These give rise to low annual temperatures for the mountain plateau and summits (Table 4.1). Table 4.1 apparently shows that there is very little difference between the three mountain summits even though Ben Nevis should be more oceanic and it is almost 100 m higher. However, the temperatures for the first three months of the year during the recording period 1985–88 were particularly low so long-term averages should probably be slightly higher for the Cairngorm locations.

The average temperatures given in Table 4.1 do not tell the whole story since there is some evidence that the corries have higher temperatures than would be expected from their altitude. This could have some large effects on the flora and fauna of the corries. In addition, although the annual average temperature is considerably lower on the mountain tops the most extreme low temperatures occur in the

Table 4.1: Mean monthly temperatures for Ben Nevis (1343 m), Ben MacDui (1256 m), Cairngorm (1245 m), Fort William (9 m) and Aviemore (229 m).

Month	*Ben Nevis*	*Ben MacDui*	*Cairngorm*	*Fort William*	*Aviemore*
J	–4.8	–5.0	–5.0	3.7	1.1
F	–4.4	–4.4	–4.9	3.8	1.2
M	–4.3	–3.9	–3.7	4.7	3.5
A	–2.1	–2.5	–1.5	7.3	5.9
M	0.7	–0.3	2.0	9.8	9.1
J	4.4	3.3	5.2	13.0	12.2
J	5.4	6.4	6.9	13.9	13.5
A	4.9	5.3	6.6	13.6	13.2
S	3.3	3.9	4.8	11.8	10.9
O	–0.3	0.0	3.2	8.1	8.2
N	–1.4	–3.1	–1.3	6.7	3.8
D	–3.5	–4.1	–2.9	4.5	2.2
Annual	–0.17	–0.6	0.6	8.4	7.1

Sources: The Ben Nevis and Fort William data are from Table 2.2, the others are from McClatchey (1996) and cover the period 1985–88

valleys during temperature inversions when cold air is trapped in the valley bottoms. For example, during one temperature inversion on 12 January 1987 the surface temperature in the Spey valley was around –30°C whilst the summit of Cairngorm was over 13°C warmer. However, the higher altitudes also experience higher windspeeds that reduce the effective temperature. McClatchley (1996) calculated that the effective temperature on the Cairngorm summit, taking account of the windchill resulting from 12 m s^{-1} wind, was around –40°C. Several ecologists have suggested that wind is one of the most important ecologically significant factors in the Cairngorms. McClatchley (1996) calculated a rate of increase in windspeed with altitude of between 9.1 and 11.0 m s^{-1} 1000 m^{-1}, which was greater than the maximum figure given by Grace and Unsworth (1988) for a range of British mountains. Extremely high windspeeds are not uncommon in the Cairngorms, the highest gust (a UK record) was 76.3 m s^{-1} (275 kph) on 20 March 1986. Gusts of over 180 kph occur most years and the average on the Cairngorm summit is 47 kph. Precipitation also increases with altitude at a rate of about 1400 mm 1000 m^{-1}. Some of this precipitation is in the form of snow, some of which can persist for long periods. It is estimated that above 914 m snow lies for an average of 165 days, rising to over 200 days above 1200 m.

Plate 4.2: Caledonian pine – remnants of a once extensive forest that covered much of the central and eastern Highlands of Scotland

The combination of long-lasting snow and appropriate sloping ground make the Cairngorms an ideal skiing location. During the Second World War the Cairngorms were used for arctic warfare training. Since 1961 the northern slopes of Cairngorm, which are just outside the boundary of the NNR, have been developed as a commercial ski resort. Inevitably this has resulted in some conflicts with conservation interests. It is not only the effects of recreation on the montane region (Chapter 3) that have been the source of concern. During the 1980s there was a plan to extend the ski resort further east into Lurcher's Gulley, an area considered by many ecologists to be the best example of an arctic-alpine environment in Britain. After a long series of legal challenges and protests the plan was abandoned. More recently there has been controversy about plans to build a funicular railway to within 150 m of Cairngorm's summit. In late 1998 a legal challenge, against the development, by the RSPB and the WWF was turned down.

It is not only the ski resort that has been the focus of environmental concerns. As with much of Scotland most of the land is divided into relatively large estates. Changes in estate ownership can have significant effects on the land management. Two of the largest Cairngorm estates, Glen Feshie and Mar Lodge, have been in the news several times during the 1990s. The WWF has described Glen Feshie as one of the worst examples of environmental degradation in Scotland. Conservation consortiums, which have included variously the RSPB, the John Muir Trust and Scottish Natural Heritage, failed to purchase this estate in 1994 and 1997. The Glen Feshie estate has passed through a series of owners in the last 30 years. It is currently owned by a Danish company. The Mar Lodge estate (310 km^2) is said by many to encompass the most important parts of the Cairngorms. The original, more extensive, Mar estate became a sporting estate in the late 1700s, its sale being part of the forfeit paid by the Earl of Mar for his part in the failed Jacobite Rising of 1715. Most of its pine trees were felled for timber, but at that time deer numbers were low and the forest was able to regenerate. However, as deer numbers rose forest regeneration was stopped while felling continued. Native pine forest now covers only 6 per cent of its former range. The Mar Lodge estate was formed in 1965 when the Mar estate was split. At that time the owner was a Swiss financier who sold the estate in 1987 to an American industrialist. In 1992 a consortium of conservation groups, including the RSPB, WWF and the John Muir Trust, attempted to purchase the estate. It is not entirely clear why their bid failed, but government (Scottish Office) policy seems to have been part of the explanation. Fortunately, the Scottish National Trust has recently purchased the estate and long required environmental improvements are now beginning, including better management of the red deer and the regeneration of important natural pine forests.

The government recognized that some action was required if the Cairngorms were to retain some of their value. They decided that one approach to the potentially conflicting interests of conservation, land owners and local people, was the establishment of a Cairngorms Working Party (CWP). The aim of the CWP was to develop an integrated approach towards future developments that attempted to achieve a consensus between interested parties. The eventual report of the working party (Cairngorms Working Party 1993) has not been received with universal approval. One of the country's most eminent mountain ecologists, who was not part of the CWP, was highly critical of the final report (Watson 1993b). There were some surprising omissions from the CWP, including relevant international experts and, perhaps most surprising, the Kincardine and Deeside District Council who are the local planning authority. It is worrying that there is no consensus for the future development of what is arguably Britain's most important wild life and wilderness region.

THE GOLDEN EAGLE

The golden eagle is one of the icons of the Scottish mountains. A sighting, albeit someway off, is usually greeted with great excitement. It has become our most impressive predator now that all of our large mammalian predators are extinct. Although the golden eagle is largely confined to the Scottish Highlands (Figure 4.3) and the Hebridean islands,

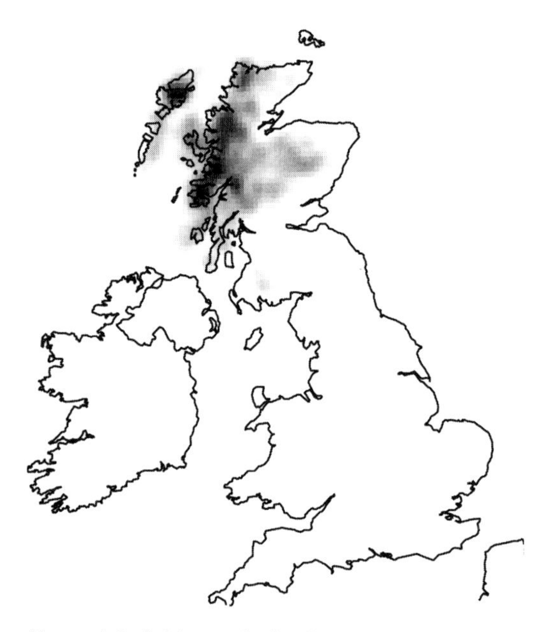

Figure 4.3: Golden eagle distribution map

breeding was once extensive in Britain and Ireland (Holloway 1996). Eagles bred in north Wales and the southern Pennines during the eighteenth century (Fryer 1987). The Welsh name for Snowdonia, Eryri, means land of Eagles. They hung on in the Lake District until the early nineteenth century (MacPherson 1892) and the southern Uplands of Scotland until the mid nineteenth century (Baxter and Rintoul 1953). Although once widespread in Ireland (Usher and Warren 1900) golden eagles were persecuted to extinction by 1913 (Barrington 1915). Currently the non-Scottish golden eagle population is restricted to a single Lake District (Haweswater) territory that is known to have produced its first young in 1970. At least one other territory that extends into northern England has been intermittently occupied, with some success, since 1970.

There is some confusion about past distribution patterns since many of the records almost certainly refer to the white-tailed eagle *Haliaeetus albicilla*. It is thought that all of the English Lake District eagles were white-tailed, and they are still recalled in place names such as Eagle Crag, Erne (Iron or Heron) Crag. This distribution of names suggests a population of about 10 pairs. There was a much more widespread bird breeding as far south as the Isle of Wight as late as 1780. There is no doubt that the white-tailed eagle was persecuted to extinction in the UK and Ireland; the last territory holding bird was shot in Shetland in 1918 (Love 1983). It seems that the golden eagle survived similar persecution because it nested in more remote areas.

Golden eagles are wide-ranging birds that are found in a surprisingly broad range of habitats. In order to be successful a pair of eagles needs access to an adequate food supply, a suitable nest site and freedom from persecution. The first systematic national survey of golden eagle numbers carried out in 1982 revealed the presence of 424 home ranges occupied by pairs and a further 87 ranges being used by single birds (Dennis *et al.* 1984). This was a substantial increase on previous estimates (Everett 1971) but it was largely attributed to improved coverage of the most remote areas. A repeat national survey in 1992 found 422 ranges occupied by pairs and 69 by single birds (Green 1996). Although the overall population appears stable there have been some marked increases and decreases between regions in the Scottish Highlands and Islands.

Large predatory and scavenging birds such as golden eagles probably gorge and fast to some extent rather than take food on a daily basis. Calculations by Brown and Watson (1964) suggested that a pair of golden eagles would require about 170 kg of meat annually. If various allowances are made for wastage (i.e. unconsumed parts of the prey), and the need to rear young and support non-breeding eagles, the total food requirements of an eagle range in the eastern Highlands rises to over

320 kg $year^{-1}$. This would be predominantly live prey (84 per cent by weight) plus some carrion. In the west of Scotland the carrion intake would be much greater. However, it is surprisingly difficult to determine what a particular golden eagle is eating. Techniques have included observation of food brought to nests and the identification of prey remains and items in regurgitated food pellets. In the United States Collopy (1983) found that there was little difference in the frequency of prey items recorded in prey remains and pellets compared with observations of prey brought to nests. However the total amount of food delivered was under recorded in pellets and prey remains. Such studies do not provide details of diet outside the breeding season or of food taken by adult birds away from the nest. In Scotland Watson *et al.* (1993) used pellets to assess diet across most of the habitats used by golden eagles. Items from pellets were placed in one of six prey categories

- deer;
- sheep and goat;
- lagomorph (rabbit and mountain hare);
- other mammals (voles, mustelids and foxes);
- red grouse and ptarmigan; and
- miscellaneous (fulmars, ducks, crows, waders, pipits, larks, thrushes and a few reptiles and amphibians).

Examination of prey items revealed marked regional and seasonal variations in the diet of golden eagles. In the west of Scotland eagles utilized a wide range of prey during the breeding season. These included hares, rabbits, grouse, seabirds, waterfowl, voles, fox cubs and occasional young otters, snakes and frogs. Many shepherds are concerned about possible lamb predation. However, it seems that, although lambs and deer calves are relatively common food items, they are usually taken as carrion. In the eastern Highlands the breeding season diet was almost entirely live prey such as red grouse, ptarmigan and mountain hares. During winter it was clear that throughout Scotland carrion from dead adult deer and sheep was consumed on a regular basis. Watson (1997) concluded that when the preferred prey (grouse and lagomorphs) are

Plate 4.3: Golden eagle nest – many years of use at a traditional site can result in a massive structure

plentiful eagles are specialist hunters with a narrow dietary breadth but that they can become generalist hunters when favoured prey is scarce.

The choice of nest site (eyrie) is strongly influenced by the need to protect eggs and young from mammalian predators including man. Proximity to foraging grounds and a desire to ameliorate the effects of weather are presumably important considerations. Golden eagle home ranges often contain several eyries that may be used in different years. The stick nests are added to each year, occasionally culminating in very substantial structures up to 5 m tall. However most nests are typically 0.5 – 1.0 m (high) by 1–1.5 m (diameter) (Watson 1997). In 1982, of 410 pairs of breeding golden eagles in Scotland, 392 used cliffs and only 18 used trees (Watson and Dennis 1992). All the trees bar one were Scots Pines situated predominantly in the eastern Highlands. Another striking feature of golden eagle nests is that they are situated at elevations approximately half that of the surrounding landscape varying from less than 200 m in the west of Scotland to 500 m in the eastern Highlands. This trait is thought to reflect the need to minimize the expenditure of energy when carrying large prey items back to the nest (Watson 1997).

The months before egg laying (November–February) have been little studied in golden eagles. Pairs of eagles are often observed engaged in high soaring and spectacular undulating flight at this time of year probably as a means of advertising their occupancy of territories. High soaring would presumably also assist in the location of carrion.

Golden eagle eggs are approximately oval (75 by 59 mm) and weigh around 145g when newly laid. Many eggs have a dull white ground colour with only faint markings but others exhibit much pigmentation with numerous brown or amber streaks and blotches. Eggs can be laid at any time between the beginning of March and the middle of April (Watson 1997). Eggs are laid at intervals of between three and five days and incubation, carried out predominantly by the female, lasts about 43 days. In Scotland most clutches are laid between the middle of March and the first few days of April but no consistent regional differences in average laying date are apparent. Watson (1997) found a significant relationship between laying date and mean temperature in February. The coldest February during the 1980s coincided with laying dates 10 days later than the much warmer year of 1990. The normal clutch size in Scotland is two eggs, although one and, more rarely, three eggs may be laid. Average clutch size in Scotland is larger in the east compared with the western Highlands and this is probably related to the more abundant live prey in the east. Throughout Scotland there appears to have been a decline in clutch size since the mid-nineteenth century and this is attributed to a reduction in preferred prey especially in western Scotland (Watson 1997). In any one year a high proportion of pairs in western Scotland fail to lay eggs and this is probably related to an insufficient availability of prey.

For about 10 days after hatching the young golden eagles are brooded by the adult female. In total the young spend about 10 weeks in the nest, eventually weighing 3.5kg (in males) or more than 4kg (females). The young birds have their maximum food requirements, which is thought to be about 2kg of prey per day, about seven to eight weeks after hatching. In many nests the larger chick kills the smaller sibling. Although the precise reasons for this Cainism are unclear it is generally assumed to be related to the amount of food brought to the nest by the parents. Young eagles spend many weeks close to the nest learning to fly and only become independent after about four months. They do not enter the breeding population until they are at least four or five years old. Knowledge of juvenile and sub adult golden eagle movements is very limited.

The rate at which raptors breed is dictated largely by the food supply (Newton 1979). Observations across Scotland have revealed that some pairs of golden eagles are consistently more successful than others in producing offspring (Everett 1971, Watson *et al.* 1989) but there are few details about which prey are consumed and how much prey is available. Watson *et al.* (1992) attempted to relate breeding success to prey availability by counting how much prey and carrion were found along line transects. The lowest levels of eagle productivity occurred in the west and northwest Highlands where mountain hares and red grouse are relatively scarce. The highest breeding success was found in the eastern Highlands where plentiful red grouse and mountain hares were associated with the more extensive heather moorland. Eagles in the Inner Hebrides and southwest Highlands also showed good productivity that was again linked to an abundance of lagomorphs.

It is also thought that weather can affect eagle performance but isolating the impact of weather on breeding success is complicated because several variables are involved and these rarely act independently. In exceptional circumstances extremes of weather can be a direct cause of breeding failure e.g. blizzards in Montana (Phillips *et al.* 1990) or too much sun in Idaho (Beecham and Kochert 1975). In Scotland cold and wet weather during late winter and early spring can adversely affect an eagle's hunting ability whilst simultaneously increasing its energy requirements. Watson (1997) observed a synchrony of breeding success in two adjacent study areas in western Scotland that had differing diets. He attributed the synchrony to the mean temperatures during February, immediately prior to egg laying. The years with milder than average February temperatures were associated with better breeding success. In an American study Steenhof *et al.* (1997) found that the proportion of eagles laying eggs, and the egg hatching dates, were related to both the severity of the winter and the number of jackrabbits.

Golden eagles have few natural enemies and, in the absence of direct persecution by man, adults probably live for between 20–30 years. Although threats to golden eagles have changed in time and space throughout its range, few golden eagle populations are entirely free from adverse actions and developments. During the nineteenth century persecution of birds of prey was intense (Love 1983). Golden eagles were poisoned, shot or trapped on most sporting estates and survived only on the most inaccessible parts of the Highlands. Two World Wars reduced the number of gamekeepers and legal protection in 1954 ensured at least the survival of the species in Scotland. However, deliberate destruction of eagles and/or their nest contents still continues today. Poisoning is the commonest method used and is most intense on land managed as grouse moor in the eastern and southern Highlands and southern uplands (Watson 1997).

Organochlorine pesticides are both toxic and very persistent in the environment. In the past they have caused great damage to raptor populations (Newton 1979). DDT and the cyclodienes aldrin, dieldrin, endrin and heptachlor are especially harmful to birds of prey. Aldrin and dieldrin breakdown to HEOD, a highly toxic chemical. DDE, the metabolite of DDT, is associated with egg thinning in raptors, leading to breakage of eggs and breeding failure. DDT was used as an insecticide in sheep dips in Britain from 1947 to the mid 1950s being replaced by dieldrin until 1966 when concern over mutton contamination prevented further use. A 10 per cent thinning in the thickness of golden eggs during 1951–65 was reported by Ratcliffe (1970). In western Scotland the frequency of egg breakage in golden eagle nests also increased at this time (Lockie and Ratcliffe 1964). Following the banning of dieldrin as a sheep dip in 1966 there was a noticeable improvement in breeding success in golden eagles in western Scotland (Lockie *et al.* 1969). The source of contamination was assumed to be the sheep carrion that was widely used by eagles for food during the winter months. More recently analysis of golden eagle eggs by Newton and Galbraith (1991) has revealed that eggshells were significantly thicker post 1970 than 1951–65. This coincided with reductions in both DDE and HEOD levels. Levels of other pollutants particularly PCBs and mercury are quite high in golden eagles inhabiting the western coast of Scotland. The source of this contamination is probably seabirds taken routinely as prey. The impact of these pollutants on coastal breeding golden eagles is unknown at present.

Since the Second World War large parts of upland Britain have been transformed by afforestation, especially the southern uplands and southwest Highlands centred on Argyll. The removal of both sheep and deer has reduced the amount of carrion available to foraging eagles. Initially, removal of sheep and deer increases the amount of ground vegetation, which in turn increases the numbers of hares and grouse, thus providing a more plentiful supply of live prey for golden eagles. However, as the forest canopy closes this temporary food bonanza is lost. In Galloway Marquiss *et al.* (1985) suggested that reduced breeding success in golden eagles was associated with extensive conifer afforestation. In Argyll Watson (1992) found a significant negative relationship between breeding success of eagles and the amount of forestry in the home range that was more than 10 years old. The effects of large scale conifer afforestation on an eagle population in any region may not become apparent for a very long time since such long lived birds can cling to territories for many years even though breeding success is greatly reduced. In the Kintyre peninsula of Argyll more than 60 per cent of the land below 200 m had become afforested by the mid 1980s, a vast increase on the 5 per cent of the 1950s. In the 1960s between 8 and 10 pairs of golden eagles bred in this area but by 1995 this figure had declined to only 4 pairs.

Low intensity pastoral landscapes are frequently utilized by golden eagles in Europe (Watson 1991). More than thirty years ago high densities of golden eagles were linked to an abundance of dead sheep and deer in western Scotland (Lockie 1964). However a long period of heavy grazing by sheep and deer in conjunction with repeated and extensive burning has resulted in the replacement of dwarf shrubs by grasses over much of western Scotland. This in turn has greatly reduced the numbers of hares and grouse that are largely dependent upon the extent and quality of heather. Eagles in western Scotland can therefore consume large quantities of carrion, occupy ground at high densities but breed rather poorly when compared with the persecuted populations in the eastern Highlands.

The effective long-term conservation of golden eagles will be achieved by a combination of measures aimed mainly at the prevention of persecution and the maintenance of landscapes with adequate supplies of food. The various strategies available include statutory protection of the species and its habitat; intervention through directed management; protection through land use incentives and constraints; and conservation education.

Golden eagles are sometimes killed or their nests destroyed by farmers and crofters who perceive a serious economic threat to their livestock. There is very little evidence to support this contention. Elsewhere golden eagles are disliked on grouse moors, not because they take some grouse as prey, but mainly because their appearance scatters the grouse making them more difficult to shoot. Education and statutory protection in the form of the Protection of Birds Act (1954) and the Wildlife and Countryside Act (1981) have already made substantial inroads into the problem. However considering the scale of raptor persecution that continues there is clearly a case for government agencies to lend greater support to the admirable campaign waged by the RSPB.

Dealing with the impacts of contemporary land use, land management and land use change is a much more complex matter. Throughout the golden eagles' range in Scotland some form of economic or social subsidy underpins most land uses. Afforestation is dependent upon substantial government grants covering the costs of establishing new woodlands. Sheep farming in the hills is supported by government 'headage' payments based upon the number of breeding ewes on individual farms and estates. Only within the statutorily designated Sites of Special Scientific Interest are proposed land use changes fully assessed for impacts on the nature conservation interest. However, many SSSIs are relatively small and only a very few encompass more than one or two golden eagle home ranges. Furthermore the total area of SSSI comprise only about 10 per cent of the golden eagle range in upland Scotland (Watson 1997). Although the golden eagle is covered by the European Union's Directive on the Conservation of Wild Birds (79/409/EEC) (Stroud *et al.* 1990) there has been little action on upland land use.

Watson (1997) argued strongly for a comprehensive strategic environmental assessment of current policy towards upland land uses. This would quantify environmental costs and benefits placing them in a social and economic context and thus highlighting areas where there is a need for change. Several areas of policy merit attention, not least the current system of sheep 'headage' payments that favour excessive stocking levels. The practice of muirburn, that is often uncontrolled in western Scotland, is ecologically damaging and needs to be discouraged. Upland afforestation policies and grant schemes also require revision to ensure that grazing animals, particularly red deer, are reduced in number when they are displaced by new forestry. This should ensure that they do not contribute to further grazing intensification. Greater financial incentives are perhaps required to encourage further development of native woodlands.

In addition, targets could be set for the expansion of breeding golden eagles into suitable areas from which they are currently excluded by persecution or distance from extant populations. Thompson *et al.* (1995a) point to the possibilities of a golden eagle population establishing itself in the northern Pennines of England. Further afield much suitable golden eagle habitat remains in the west and northwest of the Republic of Ireland. Ireland is the only country in the world from which the golden eagle has been totally eradicated. A re-introduction programme using Scottish golden eagles and the invaluable knowledge and experience gained from comparable work on sea eagles and red kites is surely a viable and highly desirable project for the beginning of the twenty-first century.

A HISTORY OF THE VEGETATION AND LAND USE IN THE SOUTH PENNINES

The south Pennines is an upland region that is close to many of the industrial towns of Lancashire and Yorkshire. Because they have experienced the effects of humans for thousands of years they provide a suitable case study to investigate, in more detail, the impact that people can have on the upland environment.

Immediately to the south of this area, the Peak District National Park forms the southern extremity of the Pennines whilst to the north lies the Yorkshire Dales National Park (Figure 4.4). Within the south Pennines a 150 km^2 tract of moorland lies to the north of the River Calder and is bounded on the west by Burnley and Colne and on the east by Keighley, Haworth and Halifax. This upland moorland area is a geologically uniform region comprising rock of the Carboniferous Millstone Grit series with small areas of coal measure shales and sandstone out-cropping in the west. Except for extensive peat deposits on the moorland plateaux, drift formations appear to be absent.

The area is dominated by land at altitudes between 250 and 430 metres with the highest point of 517 m being reached on the western extremity at Boulsworth Hill. Much of the high lying plateaux comprise quite gentle and moderate slopes (<11°) with steeper slopes of >11° and occasionally >22° where the moorland is dissected by the fast flowing headwaters of the rivers Aire and Calder. These headwaters are very deeply entrenched since they flow to a mainstream that was lowered by the deflected drainage of Rossendale during the glacial retreat phase.

During the last glaciation, although ice entirely covered the surrounding lowland, the summits of the south Pennines probably

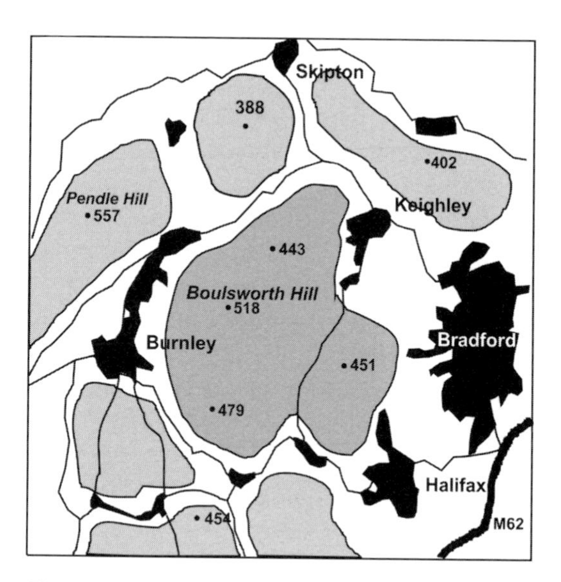

Figure 4.4: Map of the south Pennines study area

remained free of ice (Lowe and Walker 1984). However, it is possible that the sub-Arctic conditions precluded the presence of man. A few stone tools tentatively attributed to the upper Palaeolithic have been found within the study area on Midgley Moor and these could belong to this late Devensian period. There is some evidence for a short-lived but sudden climatic amelioration in Britain between 12850–12250 BP. This event, the Windermere Interstadial, has its type site in the Lake District where analysis of fossil insect assemblages indicated average summer temperatures like those of today (Coope 1977). This could have led to the utilization of the south Pennine uplands by Upper Palaeolithic man hunting game, especially during the summer months.

The widespread changes in environmental conditions at the beginning of the post glacial or Holocene period resulted in a more extensive use of the uplands by Mesolithic hunters and gatherers (Jacobi *et al.* 1976). Evidence for the presence of Mesolithic culture in the south Pennines is based on the widespread occurrence of assemblages of flint material. However these upland Mesolithic sites all occur on mineral soils with little or no associated datable material and therefore the time interval between Mesolithic occupation and the beginning of peat formation cannot be measured accurately. Moreover the subsequent development of the peat depended very much upon local topography, being earliest in hollows on the flat plateaux at higher altitudes, then spreading onto the gentler slopes and valleys at lower altitudes, and only much later invading the exposed ridges (Tallis 1964a). The only accurate date for an upland Mesolithic site in the study area is 8100 ± 150 BP for Ickornshaw Moor where a flint tranchet axe belonging to the earlier Mesolithic has been found (Faull and Moorhouse 1981). Elsewhere in the area groups of Mesolithic sites, which were probably occupied for short-term periods following hunting expeditions, have been discovered along major watersheds on Keighley, Stanbury, Wadsworth, Oxenhope, Ovenden, Thorton, Warley and Midgley Moors. Many Mesolithic sites in the South Pennines area have revealed large numbers of flint missile weapons of forms essential to a hunting economy. If the absence of large numbers of tranchet axes can be ascribed to the fact that the uplands were used mainly for hunting, then clearance of the vegetation would not have been desirable. However it is possible that some limited clearance of the moors was being made by fire and that larger scale woodland clearance was taking place on sheltered valley slopes favoured for more permanent Mesolithic settlements.

Birks (1988) and Stevenson and Birks (1995) comprehensively reviewed long-term ecological changes and the importance of climate and human interactions throughout the British uplands. Using pollen analysis the impact of grazing on heather moorland in Britain and Ireland since c.1400 is described by Stevenson and Thompson (1993). The main source of evidence for vegetation and climatic change is the pollen record preserved in the moorland peat. No detailed pollen diagrams are available from the study area and in one nearby example only, at Rishmoor Moor 15 km to the south have radiocarbon dates been obtained (Bartley 1975). Analysis of the peat profile at Rishworth Moor suggested that peat growth commenced in the Atlantic period and continued into the Middle Ages. The elm decline here was dated by radio-carbon means to 5410 ± 140 BP, slightly earlier than the generally quoted date of about 5000 BP elsewhere. The pollen consisted of some 60 per cent trees and shrubs, with the remainder being grasses, indicating an open grassy or heathy woodland environment. The predominant tree species at this time were oak, alder and hazel with some birch. The birch increased after about 4000 BP with a sporadic appearance of weeds associated with pasture and an irregular decline in the amount of oak. This overall pattern continued until about 2400 BP and may indicate slight forest clearance and the pasturing of animals. In the subsequent Iron Age, clearance of forest on a large scale, probably associated with the establishment of pastures initially, and later the introduction of cereals, is indicated by an increase in grass and weed pollen, a general decrease in tree pollen and the appearance of cereal pollen.

According to the pollen diagram farming activity continued to expand reaching a maximum around about 1920 ± 80 BP It then underwent a dramatic decline as the tree cover returned. It has been suggested tentatively that this decline in farming was associated with the Roman occupation (Bartley 1975). Near the top of the pollen diagram the evidence suggested a period of woodland clearance associated with increased pastoral and arable agriculture and this may be attributable to the late Anglo-Saxon period. Between the first century AD and the seventh century the peat profile suggested a small increase in agricultural activity but little overall change in the tree cover.

Peat accumulation requires waterlogged conditions and it has generally been assumed that the development of blanket peat on the Pennines was initiated by the onset of the wet climate of the Atlantic period about 7500 BP. Evidence from tree stumps beneath the peat, and pollen surviving in the underlying mineral soil, indicated a forest cover on the Pennines prior to the deposition of peat. It has now been suggested that the partial forest clearance by man may have created wetter ground conditions on the surface that promoted the development of peat deposits (Moore 1973). Analysis of several peat profiles has suggested that the growth of peat began at the same time as the elm decline or shortly before, and has often been linked with evidence in the pollen record of human interference with the vegetation cover (Moore 1975).

Some 15 km south west of the study area analysis of a pollen diagram from Deep Clough, Rossendale, suggested a succession of seven woodland clearance phases of limited duration, interrupted by periods of woodland regeneration (Tallis and McGuire 1972). In the Deep Clough pollen diagram dramatic changes occurred at the close of a clearance phase about 3660 BP (Tallis and McGuire 1972). Records of weed and cereal pollen ceased whilst those of tree pollen rose sharply and a widespread phase of woodland regeneration is indicated. During this period pine, elm and lime increased markedly whilst the amount of *Calluna* pollen decreased sharply. Total grass pollen remained largely unchanged and therefore it appears likely that woodland regeneration took place at higher altitudes in the *Calluna* and *Eriophorum* dominated areas. No reliable dates are available for this period of woodland regeneration and it is therefore not ascribed to any particular historical event. Tallis and McGuire concluded from the pollen evidence that despite the widespread effects on the vegetation produced by the Bronze Age clearances, subsequent climatic deterioration in the Rossendale uplands lowered human population levels sufficiently to allow substantial woodland regeneration at lower altitudes. At higher altitudes the climatic deterioration would have resulted in soil erosion and impoverishment, and the spread of peat.

Tallis (1964a and b), from a detailed examination of peat profiles from Alport Moor in the Peak District, suggested that at least three processes contributed to the current eroded mire surface. These include headward stream development (1000 BP), drying out of pools and hummocks between 800 and 700 BP and subsequent human modification of the vegetation by burning, grazing and air pollution. It is concluded however that *Sphagnum* disappeared some 400–500 BP. Tallis (1995) concluded that the presence of *Racomitrium* remains in the peat column signified a prolonged period of drier climate between 750 and 550 BP and again in the mid-eighteenth century, succeeded by a return to wetter, more humid conditions. Tallis (1997) interpreted high *Empetrum nigrum* pollen episodes prior to 2860 BP and between 800 and 750 BP as the result of extended periods of drier climate.

The evidence for the nature of the vegetation cover and vegetation change during the Middle Ages comes primarily from place names and documentary sources. The information relates mainly to woodland and woodland clearances and does not begin until after the Norman Conquest leaving a gap of several centuries when evidence is scanty. Place name evidence suggests that Angles, Danes and Norse-Irish settlers progressively penetrated the Rossendale Uplands and the South Pennines (Bartley 1967). It would seem likely that such colonization was accompanied by some woodland clearance.

Following the Norman conquest, large parts of the uplands in northwestern England were given over to hunting and also used as moorland cattle ranches, for the rearing of oxen to be sold in the markets of Wakefield, Pontefract and some Lancashire towns. The study area contained a number of cattle ranches or vaccaries during the late thirteenth and early fourteenth centuries. The moorlands were apparently being used as summer pasture and the vaccaries (or tunstalls) were in the lower valley sites.

The documentary evidence indicates the development of a pastoral economy in the Rossendale uplands. This was based on cattle and there was a gradual depletion of woodland resources during the three hundred years up to 1507. After this there was a rapid increase in settlement associated with the establishment of a more fixed farming economy, and the progressive enclosure of moorland within a system of innumerable and mostly tiny smallholdings.

Many primary settlements in the vicinity of the study area were already in existence by 1274, the date of the earliest surviving Wakefield Court roll, showing that the pattern of settlement was well established by the late thirteenth century (Moorhouse 1979). A concentrated period of clearance from the wastes or moors occurred during the years immediately before 1320, resulting in the establishment of further settlements. The late fifteenth century saw another major extension of land clearance and the foundation of new settlements (Faull and Moorhouse 1981), in a dispersed pattern of smallholdings.

Many of the field boundaries over the gritstone are now made of stone but documentary evidence (Moorhouse 1979) makes it clear that these replaced medieval hedges, probably during the late sixteenth centuries when the traditional building material changed from timber to stone. Field name evidence shows that some large fields already in existence by the thirteenth century were subsequently divided into smaller units and that some of them are known to have been cultivated. Finally, minor field-names recorded during the middle ages suggest that field barns were as common in the general vicinity of the study area as they are today in the Lake District and Yorkshire Dales (Moorhouse 1979).

In the light of documentary and palynological evidence for vegetation change it has been concluded that the cotton sedge assumed dominance over much of the blanket bog of the southern Pennines some time after the fourteenth century as a result of widespread human interference with the vegetation. These modifications, probably involving grazing and burning, resulted in an overall decline in the frequency of *Sphagnum* in the vegetation but did not cause its present virtual absence (Tallis 1964b). *Sphagnum* was certainly in abundance in the Halifax area during the seventeenth and eighteenth centuries when it was used by the mossing trade to make stone roofs watertight.

The manorial system appears to have been weakly developed and little by little individual farmers obtained the freehold of the land and this soon led to multiple inheritance and the large scale subdivision of farms. By the early eighteenth century Daniel Defoe was observing the gradual evolution of an occupational integration between subsistence agriculture and cloth making. This picture was confirmed and described in more detail by Watson (1775) and Brown (1799).

A series of Parliamentary Enclosure Acts affecting the Townships of the parish of Halifax began in 1778 and ended in 1848 producing a batch of high lying and bleak farms with their fields laid out in regular rectangles. At this time there was probably more land under cultivation for oats and potatoes than ever before and locally this was the Golden Age of agriculture (Crump 1938). The typical small holding, many of no more than 12 to 20 acres, combined subsistence farming with hand-loom weaving in a community dependent essentially on domestic industry which fed the Halifax market with cloth. It was strengthened by the abundance of yarn created by the growth of factory spinning after 1770, or a little later in the wool and worsted trades. Small spinning mills were built along the upland tributaries, a source of work for the women in the domestic weaving community; while local coal pits and quarries added to the range of male work. But this upland community saw its economic base swiftly destroyed by the spread of the power loom, from the 1840s, and the rapid development of the new factory towns of the main valleys, close to canal and railway for the cheap transport of coal, yarn and cloth. Sometime about 1870 or 1875 the little hill farm touched its zenith and no new barns or the break up of land are observed after this time. The decline in some of the upper valleys was severe as for example in Luddenden Dean where the population fell from 280 in 1855 to 96 in 1905 (Crump 1938).

Many of the small farms had a specially constructed out-house or turf cote, for storing peat cut locally on the moors and presumably peat was used as fuel for many centuries. A small number of disused peat pits can still be seen in the study area but the overall impact of peat cutting upon the flora and fauna down the ages is extremely difficult to ascertain, because once abandoned, old peat pits appear to be rapidly colonized by moorland vegetation and reabsorbed into the local landscape.

Important changes stemmed directly from the increasing industrialization from the mid-nineteenth century onwards. The advent of steam power and the burning of coal on a large scale resulted in atmospheric pollution mainly from industrial east Lancashire but also from west Yorkshire during winter (Moss 1901).

> Let us picture the view of the Lancashire towns as one stands on some of the bold western escarpments of the Pennines, say from Blackstone Edge or Gorple Stones. Deadly suburban fields form the most extensive element of the background; but what rivet the eye are the scores, and scores again, of mill chimneys, tall straight, and lank, belching forth volumes of black, dense smoke straight at the rocks on which we stand. Rochdale, Littleborough, Bacup, Burnley, Nelson, Colne each contributes its quota, for from them, i.e., from the west, the wind blows for nine months out of the twelve: and during the remaining three, Huddersfield, Halifax, Sowerby Bridge, Hebden Bridge, Keighley, Bradford, effectively maintain a continuous supply. Swill Hill is well-nigh permanently embedded in a smoke cloud, being only three and half miles south-east of Manningham Mills, Bradford, on the one hand, and the same distance north of Dean Clough Mills, Halifax, on the other. Further, given a clean expanse of white snow to begin with, and a driving south west drizzle to follow, and even on the remotest moors of Halifax, the whole of this clean sheet is in two or three hours palpably blackened.

Nowell (1866) connected the superabundance of factory smoke with the disappearance of a considerable number of mosses from the vicinity of Todmorden that forms the southern boundary of the study area. The impact of 'acid rain' in the southern Pennines is reviewed by Press *et al.* (1983) who, like Tallis (1964b), concluded that the main effect of this pollution upon the vegetation cover appears to have been the almost total elimination of *Sphagnum*.

With industrialization came an expanding population that required a pure public water supply. This resulted in the construction of reservoirs, particularly at the heads of cloughs. Ancillary works associated with this often included the draining of the moors themselves and the deflection of stream systems to divert water into the new reservoirs, thus greatly changing drainage conditions. At the same time the Water Corporations completely removed cattle from the gathering grounds and severely restricted the number of grazing sheep. The period of reservoir construction lasted sixty years from about 1875 to 1945 and several upland farms were lost in the process and reverted to moorland.

Moss (1900) commented that even on the tops of the moors remotest from the town, destructive effects may be noticed due to smoke largely brought over from Burnley and Rochdale by the prevailing west winds.

> The chief sufferers however, among the moorland plants are those which inhabit wet places. Their disappearance is to be accounted for by the drainage of the moors, by the making of reservoirs at the heads of the cloughs, by the cultivation of heather to provide cover for the grouse, and by the tilling of the ground at the moor edges.

These activities resulted in the disappearance of plants such as the bog pimpernel *Anagallis tenella*, the lesser twayblade *Listera cordata*, bog myrtle *Myrica gale*, juniper and three clubmosses *Lycopodium* spp. (Moss 1900). The later phases of industrialization and population growth also coincided with increasing interest in grouse shooting and the large-scale management of the moors for grouse. The associated muirburn is thought to have been responsible for the disappearance of some already rare moorland plants (Moss 1901). Finally, Moss (1900) referred to the depredations of pedestrians, herbalists, gardeners and even naturalists whose activities culminated in the disappearance of ferns, or plants with showy flowers, or plants of some rarity such as bog rosemary *Andromeda polifolia*.

Overall the pattern of land ownership and land use has been relatively stable during the twentieth century, the major incursions being the construction of reservoirs, at Walshaw and Gorple in the 1930s and 1940s. Since the Second World War there has been, until quite recently, a steady decline in grouse shooting as witnessed by a decrease in the number of gamekeepers employed on the moors; today only four full-time and two part-time keepers remain. Forestry has had minimal impact in this area as yet and is largely confined to a series of small trial blocks. It has often been claimed that the main factor limiting extensive commercial afforestation in the past was airborne pollution. This is still felt to be a serious problem, limiting the range of species which can be grown and retarding the growth of those that can be established (Lines 1984).

BRACKEN

There are two plants that many consider to be weeds of upland habitats. Rhododendron *Rhododendron ponticum* is a localized problem in parts of Snowdonia. Because it is a non-native garden escape it has few natural enemies and British upland habitat is not dissimilar to its native habitat (Thomson *et al.* 1993). The second weed, bracken *Pteridium aquilinum*, is a much more serious problem.

Bracken is an especially widespread fern occurring on every continent except Antarctica (Page 1979). Within Britain it is abundant in western Scotland, northern and western England and Wales. It is probably a native plant of woodlands especially glades and woodland edges, but it is also well suited to life outside woodlands. Estimates of the amount of land covered by bracken vary widely. Taylor (1986) gives a figure of 6361 km^2 or 2.8 per cent of Great Britain whilst Bunce and Barr (1988) quote 3000 km^2 or 1 per cent which is close to the estimate of 3613 km^2 obtained from the Land Cover Map of Britain (Fuller *et al.* 1994). These variations arise because there are difficulties mapping bracken from aerial photographs and satellite images. In addition there are problems identifying boundaries of bracken dominated land and extrapolating from detailed local studies to the regional or national scale.

Bracken can be an important breeding habitat for moorland birds particularly whinchat (Allan 1995) and other species including ring ouzel, hen harrier, merlin and twite (Haworth and Thompson 1990). Some rare plants are found associated with bracken including the autumn crocus *Colchicum autumnale* and Solomon's seal *Polygonatum multiflorum* (Pakeman and Marrs 1992). Two species of butterfly, heath fritillary *Mellicta athalia* and high brown fritillary *Argynnis adippe*, use food plants present under light bracken cover. Overall bracken is known to support over 40 species of invertebrate (Lawton 1976) and some reptiles and mammals can benefit from the shelter it provides. It is also considered to be an important characteristic of the upland landscape during autumn and winter.

Unfortunately bracken is also a vigorous and aggressive fern that can spread rapidly by means of strong underground rhizomes. Rates of bracken invasion range from 1 to 3 m per year. In general most bracken areas are considered less important from a conservation point of view than the plant communities they have replaced. Estimates for the amount of new ground colonized each year vary between 1 and 3 per cent (Taylor 1986, Hopkins *et al.* 1988, Miller *et al.* 1990, Pakeman and Marrs 1992). Marrs and Pakeman (1995) suggest that global warming could encourage additional bracken expansion and may make the plant more difficult to control where it occurs.

Bracken causes economic loss to farmers because it reduces grazing and increases animal welfare expenditure because bracken can be poisonous and it increases the risk of tick-borne diseases. Evans (1986) and Taylor (1989) review these problems extensively. Sheep ticks can be abundant in bracken litter and may carry diseases such as louping ill that affects both sheep and grouse (Hudson 1986b). Ticks may also carry Lyme disease that affects humans. In some areas higher levels of certain cancers have been linked to bracken (Galpin *et al.* 1990).

Bracken is also able to compete with and damage young trees and is now known to have considerable effect on hydrology by intercepting up to 50 per cent of incident rainfall (Williams *et al.* 1987). In some locations it is also necessary to protect archaeological features from damage caused by bracken rhizomes.

During summer 20–60 bracken fronds per m^2 can develop with a mean height of between 1–2.5 m (Marrs and Pakeman 1995). The resulting deep shade excludes other species. Dead fronds lying on the ground in autumn contribute a substantial amount of litter and this may prevent germination and establishment of other plant species. Underground, bracken rhizomes contain reserves of carbohydrate, nutrients and frond buds. Initially new fronds emerge and utilize the carbohydrate reserves in the rhizomes. Later in the season the growing fronds transport carbohydrates and nutrients back to the rhizomes.

Bracken is difficult to control because it is not very susceptible to damage. Rhizomes contain active buds that are not used in any particular year and therefore new fronds can always be grown to replace damaged ones. In addition all parts of the bracken plant possess anti-herbivore defences including cyanide (Hadfield and Dyer 1986), thiaminase (Evans 1986), phenolics and tannins. Ecdysones, capable of disrupting insect development, are also present and bracken may also produce allelopathic chemicals (Gliessman 1976).

Bracken grows best in deep well-drained soils and is limited by waterlogged conditions or thin soil (Watt 1976, 1979). It is possible that moorland drainage operations have assisted the spread of bracken. The fronds are especially sensitive to frost and this is the main factor limiting altitudinal distribution (Watt 1976). Other vegetation, the rate and type of animal stocking and insect herbivory are all thought to limit the growth of bracken. However, apart from Watt's (1955) famous example, evidence for vegetation suppressing the growth and development of bracken is very limited. Watt (1955) described a lowland heath situation where bracken invaded *Calluna* stands only when *Calluna* was in its pioneer and degenerate phases. When *Calluna* was competitively strong in the building and mature phases bracken retreated. This heath has, however, now been completely invaded by bracken (Marrs and Hicks 1986). Anecdotal evidence suggests that changes in upland land management from cattle to sheep and wether sheep to ewes and lambs has assisted bracken expansion by reducing the trampling pressure that is known to damage fronds. Although approximately forty species of insect feed on bracken, they rarely seriously damage the plant because it has herbivore defence compounds and insect predators and pathogens (Lawton *et al.* 1986).

Roberts *et al.* (1996) describe the physical and chemical control methods that are the two main approaches to bracken control. In the autumn, before any bracken control starts, cutting or burning can disturb the litter at the bracken site. This exposes the rhizomes and allows frost to penetrate (Gimingham 1992). It also improves access and may allow some early vegetation regeneration. Physical control also involves the cutting or crushing of growing fronds so that the surviving rhizomes are gradually starved (Williams and Foley 1976). This involves a long-term approach but has the potential advantage of lower cost and is less dependent on weather conditions than chemical control. Furthermore it does not damage non-target plant species.

Because rolling and cutting are threats to ground-nesting birds they cannot be used during the nesting and fledging period. Indeed it may be advantageous to retain some bracken for its value as a nesting habitat. Cultivation is also an effective treatment on areas accessible to machinery. This exposes the bracken rhizomes to winter frost. Some control can also be achieved by the short-term use of stock on areas of bracken to break up the litter, again allowing frost to damage the rhizomes. Winter feeding is a vital part of this method as stock may otherwise be poisoned by eating dead bracken or rhizomes. Bracken is also sensitive to trampling during early periods of frond growth. However stock may need to be moved in the spring to stop them eating young bracken. Bracken cutting, repeated for at least three years, can be used on more mature fronds which need to be cut twice a year (Williams 1980).

Although burning bracken litter may be helpful in providing access and easing cultivation it serves no direct control purpose, except that autumn burning may help with frost penetration of the rhizome. Where muirburn takes place in bracken, experience has shown that there may be two consequences. Either the bracken will invade heather because competition has been reduced, or the heather will establish on the bracken site, helped by the removal of bracken litter and exposure of bracken rhizomes to frost damage. Establishing tree-cover can, in the long term, suppress bracken growth by shading. The initial establishment of trees may be difficult because of competition from bracken. However, weed control is an important part of most woodland establishment.

Two translocated herbicides can be used for chemical control. They are asulam (sold as Asulox) and glyphosate (generally sold as Roundup). Asulam is the most widely used because it is cheaper and more specific than glyphosate. Compared with most herbicides, asulam is reasonably specific, principally killing ferns (Veerasekaran *et al.* 1976, 1977a, b, 1978). However, some effects of asulam have been noted in non-target species. Because glyphosate is a broad-spectrum herbicide that affects most forms of vegetation it cannot be used where it may affect any non-target species. Glyphosate is only suitable where other vegetation is absent, for example in tracts of dense bracken with thick litter and no understory of grass or heather. Glyphosate has some advantages. It has a wider window of application than asulam and it produces browning symptoms, so that the evenness of application can be judged in the year of spraying. Timing of the application is important for effective results. Maximum absorption and translocation by the rhizome is achieved by spraying when the fronds are fully expanded and bright green, and before any dieback occurs. This is usually mid July to late August, depending on the altitude and season.

Without effective aftercare, bracken will stage a rapid comeback (Robinson 1986). Regenerating fronds or areas missed during initial control must be brought under control. Follow-up may be by chemical or physical control, and a combination of different methods can be beneficial. For example, a single cut of bracken can create an even canopy, a higher density of fronds and more active buds on the rhizome. This increases the efficacy of herbicide that can then be applied in the following year. In general, two years must be allowed between phases of spraying in order to allow dormant buds on the surviving rhizomes to emerge. Trampling by stock can also help to suppress surviving fronds on sprayed areas other than on sites sensitive to trampling, e.g. archaeological sites. Young bracken fronds growing just below the surface are particularly sensitive to treading during the spring. Cattle are more effective than sheep but potentially cause more damage to other vegetation. Stock treading also increases the rate of breakdown of dense bracken litter.

Depending on the long-term objectives, treated areas may be subsequently managed in a number of ways. Grass seeds (e.g. bent or fescue) or heather cuttings can be used to encourage new vegetation. This can be assisted by the application of small quantities of fertilizer and a period when no grazing is allowed. Periodic adjustment of stocking rates can be used to allow the development of desirable vegetation. In some areas vigorous heather regeneration may keep bracken in check in the future, but stocking at too high a density, even for short periods will hinder this effect.

5

PRACTICAL WORK

•

By far the best way of understanding the ecology of a habitat is to investigate it for yourself. In order to gain maximum benefit from ecological studies it is important to plan in advance. Part of the planning is to be sure of the questions being asked or problems to be solved; the most elegant research tends to ask simple questions or look at simple problems. Prior planning also includes identifying the methods for data analysis (see below). Equipment should be checked and the experimenter should be fully competent in its use before gathering data. A consistent approach, using the same techniques and only varying the factor to be analysed is usually the best approach. Often a small pilot study will help the main investigation to run smoothly and will allow the methods to be refined.

It is important that permission is received from the owner(s) of land being used for the project, both to gain access and to carry out particular types of experiment. For example, it is against the law (Wildlife and Countryside Act, 1981) for any unauthorised person (i.e. anyone who is not the owner, occupier or has been authorized by the owner or occupier of the land concerned) to uproot any wild plant. Other organisms are even more strictly protected; almost all birds and some other animals, together with several plant species, are fully protected under the law (see Jones 1991, for further details). Care should be taken to comply with the law, to ensure that you do not damage or unduly disturb any plant or animal, and to design and implement any research project so as to leave the habitat as it was found.

Upland systems provide opportunities for a wide variety of ecological research. The following projects have been selected because they can be investigated with relative ease, require little in the way of equipment and utilize common habitats or species. They are designed to ask simple questions and provide data to analyse with simple tests. As in most research, the results obtained may stimulate further work on supplementary questions.

EXPERIMENTAL DESIGN

Comparisons between situations in which only one factor varies are the easiest to interpret. For example, comparisons between several sites that differ in size, but are similar in all other respects will allow an examination of the influence of size of site on whatever is being recorded. The experimental design must be considered in some detail. There are two major types of experimental design: observational and manipulative. The majority of projects described below are observational. Here, a variable (the behaviour of animals, the percentage cover of different plant species, the numbers of animals of a particular species, etc.) is recorded under different circumstances (different sites, weather conditions, altitudes).

Analysis is often a matter of finding whether the variable measured differs in two or more circumstances (e.g. sites, times of day). If two variables (e.g. number of species and temperature) are measured, you may wish to examine the relationship (called correlation) between the two. Two variables may have a positive (as one increases, so does the other), negative (as one increases, the other decreases) or no relationship. However, with observational experiments you will not be able to say definitively that a change in one variable causes the change in the other. This is because an observed increase in one variable (e.g. the number of a species) may be correlated with a measured rise in another factor (e.g. temperature) but actually be due to changes in a third, unmeasured variable (e.g. the abundance of its prey). In manipulative experiments, on the other hand, the experiment is designed such that one variable is altered by the experimenter (pH, temperature, fertilizer concentration) allowing a much greater emphasis on cause and effect when it comes to analysing the data. However, since these are often conducted in fairly artificial conditions, they may have less relevance to real world situations than may be achieved with observational experiments.

It is important to take several replicate samples, since data gathered from only a single sample may not be representative of the situation in general. In order to reduce bias, sampling should be systematic or random. It is essential that data gathered are recorded correctly to avoid later problems in analysis and interpretation. Record your data in a hard-backed book rather than loose sheets of paper to reduce the risk of loss. Where data are to be analysed using a computer, it is useful to transcribe the data onto an appropriate spreadsheet or other data file as soon as possible, and to check the transcription carefully to identify and rectify any mistakes. The techniques which could be used in the analysis of such data are beyond the scope of this book, although methods of analysis are suggested in the project descriptions, and several texts which provide further details are listed in the further reading section.

HEALTH AND SAFETY

Health and safety in the field (and laboratory work) should be paramount. Do not engage in behaviour or activities that could harm yourself or others. Assess the risks and health and safety issues which are likely to be involved and to protect against them. General safety issues include: wearing appropriate clothing for the time of year and terrain; using safe equipment (e.g. plastic tubes rather than glass); not working far from help (ideally work in groups of two or more); informing someone responsible of the details of planned field work in advance (including location and duration) and 'signing off' with that person on return.

Soil and water contain organisms and compounds which are hazardous to health. Gloves should be worn for all fieldwork involving these materials, and when handling spiny, or otherwise hazardous plants. Tetanus is a potential hazard for anyone working out of doors, especially those in contact with soil. Spores of the tetanus bacillus live in soil, and minor scratches (e.g. from bramble thorns) could provide a point of infection. Immunization is the only safe protection and, being readily available, should be kept up to date. Weil's disease (or Leptospirosis), is caused by a bacteria carried by rodents (especially rats). Urine from infected animals contaminates freshwater and associated damp habitats such as river, stream and canal banks, and is more common in stagnant conditions and during warmer months. Infection is usually via cuts or grazes, or through the nose, mouth or eye membranes, and precautions should be taken to avoid contact between these areas and potentially infected water. Cover cuts and grazes with waterproof dressings, use appropriate waterproof clothing including strong gloves and footwear, and avoid eating, drinking or smoking near possible sources of infection. Lyme disease is another potential hazard for fieldworkers. This is transmitted by female ticks which, although more usually a problem in uplands and woodlands, have been found in urban parks. They are especially likely to bite from early spring to late summer. To help avoid the disease, prevent ticks from biting by wearing appropriate clothing (e.g. long trousers), check for the presence of ticks (light coloured clothing helps) and remove ticks as soon as possible if bitten (twist slowly in an anti-clockwise direction without pulling and seek medical help if mouthparts remain within the skin).

In addition, it must be remembered that upland habitats are potentially dangerous places. In particular you must consult weather forecasts before venturing into the uplands and you must ensure that you have adequate clothing, equipment and food. Weather conditions can change rapidly and conditions at low level may not be representative of what is happening at the top of a mountain. Map reading skills are essential for work in the uplands. It is always wise to refer to the map continuously. It can be very difficult to work out where you are now if you did not know where you were previously! The safety issues indicated here are not comprehensive. Before any fieldwork is undertaken you are advised to consult appropriate publications such as that produced by the Institute of Biology (Nichols 1990).

SUGGESTED PRACTICAL WORK

HEATH RUSH SEED HEADS

Aim: to investigate how a common plant responds to the changing environmental conditions that are associated with increasing altitude.

In some early work Pearsall (1971) found that the number of flowers, the length of the flower stalk and the number of mature capsules of heath rush plants varied with altitude. Details are provided in the Species Box 2.14. Since this example illustrates how a single species adapts to changes in the environment associated with increasing altitude it is worth repeating the work.

Data must be collected in late summer or early autumn when seeds should have been produced. You should aim to collect data from plants whose altitude extends over at least 300 m. At each sample location (these will be determined by the plant's local distribution) you should record the flower height, the number of flowers per plant and the number of seed capsules per plant. The first measurement can be obtained from a large number of plants (>25). The last two will take much longer, so it will be necessary to reduce the number of replicates. You should compare how these variables relate to the altitude at which the plant was growing.

GROWTH AND GRAZING EXPERIMENTS ON PURPLE MOOR GRASS

Aim: to investigate, under field and laboratory conditions, how a common upland plant responds to processes such as grazing and soil nutrient status.

Purple moor grass is a particularly useful plant for exploring several aspects of upland ecology. Data can be collected from different locations and used to investigate the amount, and potential causes, of any variability. It is also possible to use this plant for laboratory experimentation.

Field data collection and analysis

Using tussocks obtained from different locations separate out at least 50 internodes. Cut all roots near to the base of the internode but ensure that the small new shoots and buds are not damaged. For each internode record:

- the weight of the internode and its new shoots;
- the total number of new shoots;
- the total length of all new shoots;
- the length of the internode.

You should analyse your data, e.g. calculate means and standard deviations and plot suitable scatter diagrams and then attempt to answer the following questions.

- Are there any differences between your samples?
- If there are differences can they be related to the environmental conditions at the sampling sites?

Some example data are presented in Table 5.1 for two samples obtained from two sites in Lyme Park, Cheshire. Sample A comes from an exposed location where there was little grazing. Sample B comes from a more sheltered location where there was some grazing by red deer.

Laboratory studies

Plants can be grown from internodes (Species Box 2.13) under a range of experimental conditions. Ideally these experiments should be carried out between November and March when the plant is 'dormant'.

A single tussock of purple moor grass will usually provide 50–100 internodes. Individual internodes can be excised from the tussock and planted in a sand/loam mixture. Before planting all but the longest bud should be removed from each internode and the total roots trimmed to a common total length (e.g. 2 cm). The internode weights and lengths should be recorded before planting. Only the bases of the internodes should be covered by soil.

It is possible to design many experiments using this simple system (Latusek 1983). For example what are the effects of grazing, soil fertility, soil pH on plant growth? Do plants from different locations show similar responses?

Simulated grazing

How does a grass respond to different levels of grazing? It is much simpler to manipulate the conditions in a laboratory than in the field.

Sufficient internodes should be obtained to allow adequate replication (at least 10 replicates per treatment). You should also think about the layout of the plant pots and the effect of any uncontrolled environmental gradients such as light and temperature.

Once sufficient leaf growth has occurred the plants can be defoliated artificially using scissors. Some possible treatments are listed in Table 5.2.

It is suggested that maximum shoot length and total shoot length are recorded weekly and that these are used to calculate a relative rate of shoot elongation (RRSE = [length_t–

Table 5.1: Example data collected from two purple moor grass populations in Lyme Park, Cheshire

	Sample A frequency	*Sample B frequency*		*Sample A frequency*	*Sample B frequency*
Number of shoots			**Internode length (mm)**		
0	6	1	11–20	0	1
1	49	13	21–30	31	11
2	107	29	31–40	75	47
3	33	6	41–50	98	67
4	25	1	51–60	82	21
5	0	0	61–70	32	6
Total shoot length (mm)			**Internode weight (g)**		
11–20	15	1	0.00–0.10	4	0
21–30	21	3	0.11–0.20	24	15
31–40	39	6	0.21–0.30	51	26
41–50	35	9	0.31–0.40	56	38
51–60	34	8	0.41–0.50	49	34
61–70	27	18	0.51–0.60	29	14
71–80	38	9	0.61–0.70	11	0
81–90	18	17	0.71–0.80	5	3
91–100	15	13	0.81–0.90	6	0
101–110	15	16	0.91–1.00	0	0
111–120	11	13	1.01–1.10	1	0
121–130	5	8	1.11–1.20	1	0
131–140	7	13	1.21–1.30	1	0
141–150	1	4	1.31–1.40	0	0
151–160	2	2	1.41–1.50	1	0
161–170	0	1	1.51–1.60	0	0
171–180	0	2			
181–190	0	0			
191–200	1	2			

length_{t-1}]/length_{t-1}, where length_t is the current length and length_{t-1} is the previous length). Graphs can be plotted of RRSE against time.

This experiment can be continued until the plants have flowered and produced seed.

Soil fertility

Use the same method as above for the preparation of the internodes. Plants can be grown in a variety of potting soils, with and without the addition of soil fertilizers. Similar measurements of plant growth can also be made to

Table 5.2: Possible simulated grazing treatments for purple moor grass

Control:	no defoliation
100 per cent defoliation:	once (6 weeks) towards the end of the experiment
100 per cent defoliation:	once (3 weeks) towards the beginning of the experiment
50 per cent defoliation:	once (6 weeks) towards the end of the experiment
50 per cent defoliation:	once (3 weeks) towards the beginning of the experiment
Weekly defoliation:	leaving a constant amount of leaf material, e.g. 1 cm

investigate how the plants respond to soil conditions.

FOOTPATH EROSION

Aim: to develop a methodology that can be used to assess the amount of damage associated with footpath usage.

Footpath erosion can have quite extensive effects. For example, deep gullies may be formed and the soil may be washed away in quite large quantities. Alternatively soil may be compacted or, if the soil conditions are wet, the path can become very wide as walkers attempt to avoid getting their feet wet. It is often desirable or necessary to introduce management techniques that will minimize such damage. However, before possible management options can be considered it is necessary to have a standard method for assessing the condition of a footpath.

You will need to measure the physical and biological characteristics of the path and its surroundings. Once a suitable location has been identified the characteristics of the path can be determined from transects that run at right angles to the direction of the path. For example, slope is probably an important determinant of gully formation and some species such as crowberry and cotton-grass are known to be susceptible to trampling. A 1971 Countryside Commission report calculated an Index of Extent (IE) footpath statistic that took into account the variation across the full width of a path. IE = (path width + bare ground width) – (undamaged vegetation width). Units are in metres since all are distance measures (across a cross-section). On a busy path such as the Pennine Way IE values greater than 100 m were recorded.

It is suggested that you measure path width, path depth, slope, aspect, soil depth/soil compaction, vegetation (species, cover, height), bare ground across several transects that include sloping and flat areas if available.

You should consider how walkers, horses and bikes (pedal and motorcycle) may have different effects and think about ways of controlling and repairing the damage that they cause.

See Grieve *et al.* (1995) and Pearce-Higgins and Yalden (1997) for additional information.

SHEEP BEHAVIOUR

Aim: to develop an understanding of the methods that can be used to record the behaviour of an animal and to assess how such behaviours can affect its impact on the vegetation.

No study of the uplands would be complete without some investigation into the relationship between sheep and vegetation. The social structure of sheep appears to be quite important in determining where individual sheep will graze. In addition to the obvious effects on the sheep, this may also have consequences for the vegetation. Although sheep are social animals, the individuals within a flock tend to form groups that maintain their own 'home range'. In northern Britain the home range is known as a heft. Intruders from other groups are repelled. The size of a group can vary considerably, partly due to the number of surviving lambs. Grubb (1974) suggests that these groups are restricted to related individuals. This group structure has important implications for grazing management since increasing or decreasing the number of sheep may not affect the grazing pressure on a particular area (Yalden 1981). It has been suggested that sheep move down the hill during the evening and back up the hill the following morning. This movement may also be related to the timing of grazing, which appears to be mainly during the day (Grubb and Jewell 1974).

Before the behaviour of sheep can be recorded you need to know something about their behavioural repertoire (ethogram). Individual behaviours must be clearly and unambiguously described before recording begins. This usually involves a period of preliminary observations. Once you are familiar with the range of behaviours you should consider constructing a recording sheet. For the purposes of this exercise three behavioural categories may be sufficient (Table 5.3).

One of the questions that you could attempt to answer is 'What is the structure of the feeding behaviour?' There are several ways of obtaining suitable information.

- When does feeding begin?
- How long is a feeding bout? (Bouts are separated by some arbitrary time interval.)
- Is feeding coordinated? (Do they all feed at the same time, on the same vegetation?)
- Does biting rate vary with forage quality? (This can be sampled over small times within a bout.)
- Is the biting rate linked to the step rate? (These determine how quickly sheep progress through a habitat.)

Table 5.3: Possible behavioural categories for recording sheep behaviour

Category	*Notes*
Feeding	Where?
	When?
	On what?
	How long is a feeding bout?
	Biting rate?
Movement	Direction
	Distance
	Duration
	Why?
Interactions	With other sheep, e.g. what is the distance to the nearest individual?
	Frequency of vocalizations
	Others (e.g. with people, crows), how is vigilance shared between individuals?

- Do they move continuously or intermittently? (What are the intervals between moves, how far do they move?)

In addition you could examine aspects of group structure. For example:

- How many groups can you recognize, are they the same size, are they permanent?
- How are the groups dispersed over the range?
- Is vigilance (head up) related to group size, is vigilance shared?
- How important are vocalizations to group structure?

Behavioural data could be obtained from video recordings. In this way regular samples can be obtained over a long period by fast forwarding the tape. You should consider both focal (concentrate on known individuals) and scan (consider the whole group) sampling techniques.

More detailed aspects of sheep behaviour can be found in Lynch *et al.* (1992).

COMPARATIVE STUDIES ON BRACKEN

Aim: to investigate factors that influence the growth and productivity of bracken.

Bracken is perceived by many to be a problem plant in the uplands. Understanding the conditions that promote bracken growth may help us to understand which areas are under threat from potential bracken invasion.

There are a number of factors that are thought to influence the performance of bracken. If comparisons are to be made between different stands it is important that they are at the same development stage. Fronds begin to grow in early May. The canopy becomes fully established in late July and senescence begins towards the end of August (Nicholson and Paterson 1976). The size of the frond also appears to be related to its position within the bracken community. For example, fronds at the invading edges tend to be smaller (Watt 1976).

Transects (minimum width 1 m) can be established that penetrate from the edge of the stand towards its centre. At regular intervals the density, size and standing crop of bracken fronds should be measured. Since bracken expansion is dependent upon sufficiently deep soil it is probably worth attempting to measure soil depth. Results from several bracken patches can be compared.

There is much scope for extended projects involving bracken. In particular changes in bracken chemistry (Lawton 1976, Watt 1976, Williams and Foley 1976) and the structure of the invertebrate community (Lawton 1976) are likely to be worthwhile.

Note that bracken litter may contain sheep ticks. Wearing gaiters may help to protect the legs. It is advisable to inspect the limbs for evidence of ticks at the end of the day.

RECORDING VEGETATION CHANGE USING AERIAL PHOTOGRAPHS

Aim: to demonstrate the value of aerial photographs as historical records of habitat conditions.

Aerial photographs can be used to obtain important historical information about the condition and extent of vegetation over relatively long periods of time. They are particularly useful in regions that are thought to have experienced significant change, including afforestation and changes in grazing or recreational pressure. The photographs can be used in conjunction with Ordnance Survey maps to trace changes in forest cover. Some suggested regions for future study are listed in Table 5.4.

Table 5.4: Possible regions for aerial photographic monitoring

Region	*Trends*
Kinder Scout (Peak District)	Increased recreation, erosion of peat and changes in grazing pressure
The Lui, Derry and Quoich Glens near Braemar	Loss of native pine forest, new forestry plantations
Cairngorm or GlenShee (Cairnwell)	Skiing and recreation (new footpaths)
Any region with grouse moors	Changes in the extent of managed moor and patch size statistics

A brief introduction to remote sensing, including mapping from aerial photographs can be found in Budd (1991). Although the interpretation of aerial photographs is a skill that requires considerable training it is possible for the novice to obtain an insight into habitat changes that have occurred between surveys. It is important to understand that aerial photographs are not maps. This is because in their unprocessed state they are not geometrically accurate. The science of photogrammetry is concerned with the correction of topographical distortions that are present in all aerial photographs. However, for the purposes of this exercise such distortions will be ignored.

The visual interpretation of an aerial photograph can be very subjective. There is no doubt that familiarity with the location greatly improves and speeds the process. Although the two most useful properties for delineating blocks of habitat are colour and texture, there are other 'clues' that could be used. Tone (hue or colour) refers to the relative brightness or colour of regions and is the most important feature on a photograph because without it different elements could not be recognized. The texture (smoothness or roughness) of the photograph's features is related to amount of tonal change. Texture is produced by features that are too small to identify. For example even aged stands of heather may seem smooth in comparison to a forest canopy. An awareness of scale can help to identify some objects, for example is a dark patch a puddle, pond or lake? A regular outline or pattern usually indicates an anthropogenic feature. Some objects, such as cultivated fields or forest plantations, can be identified almost solely on the basis of their shapes.

Sources of aerial photographs

Recent photographs (after 1980)

The Ordnance Survey has a good supply of recent photographs. Details can be obtained from their Air Photo Sales department (Phone 01703 792584, Fax 01703 792250). If you are able to provide a geographic name, or a grid reference, for the area you are interested in they will supply a flight diagram. The flight diagram shows details of their aerial coverage of that area. The centre points of photographs are marked and, using a tracing paper template overlaid on the diagram, you should be able to work out the area covered by each photograph.

Earlier (e.g. 1970s)

Photographs of Scotland are available from
The Royal Commission on the Ancient and Historical Monuments of Scotland (RCAHMS)
John Sinclair House,
16 Bernard Terrace,
Edinburgh, EH8 9NX

Tel: 0131–661–2278
Fax: 0131–662–1477/1499
E-mail *rcahms.jsh@gtnet.gov.uk*

Elsewhere
The Royal Commission on the Historical Monuments of England (RCHME),
National Monuments Record Centre,
Kemble Drive,
Swindon SN2 2GZ
Telephone : 01793 414733
Fax : 01793 414606
E-mail *info@rchme.gov.uk*

They have two collections. The Vertical Collection is probably the most useful for vegetation surveys. This consists of over 3 million air photographs, including the national cover taken by the RAF in the 1940s. Photographs are available from three periods.
1940–65, including the 1946/7 national survey at a 1:10,000 scale, others are at 1:2,500 – 1:60,000 (RAF)
1952–79. 1:5,000 – 1:23,000 scales (Ordnance Survey)
1952–84. 1:3,000 – 1:30,000 scales (Meridian Airmaps Ltd)

Requests need to be written (letters, faxes, e-mails or completing a form on their Web site http://www.rchme.gov.uk/air-help.html). As with the Ordnance Survey you must provide details of the area of interest in the form of an Ordnance Survey grid reference and, where possible, a photocopy of an OS map with the area marked on it. They may also be able to help to identify other sources of air photography.

GLOSSARY

•

Acid deposition The burning of fossil fuels results in the emission of sulphur and nitrogen oxides that increase the acidity of rain and snow ('acid rain'). The pollutants can travel great distances before being deposited; hence the amount of acid deposition also depends on the local and regional weather patterns. Its effects are exacerbated where soils are thin and acidic.

Acid soil A soil that has a pH less than 7.0 or more usefully a pH less than 6.6. Soil acidity is a consequence of the presence of many hydrogen and aluminium ions. Important plant nutrients tend to become less available as soil acidity increases.

Albedo A measure of surface reflectivity, strictly the ratio of light reflected by the earth's surface to that received from the sun.

Alpine A montane environment characterized by Central European mountains that experience a continental climate.

Anthropogenic An effect resulting from the actions of people.

Arctic Treeless tundra that is north of the polar tree line.

Base-deficient Soils that have a shortage of cations such as calcium and magnesium. Consequently the dominant cations tend to be hydrogen and aluminium, resulting in soils that are quite acidic.

Boreal A region of coniferous woodland typically associated with snowy winters and short summers in northern Scandinavia and Russia.

Boreo-alpine A biogeographical term that refers to organisms who have disjunct distributions associated with mountain ranges in central and western Europe.

BP Before Present, an alternative to BC and AD.

Bryophyte Plants (mosses and liverworts) that lack true vascular tissue and roots. Reproduction is via spores.

Calcifuge A plant that grows best in acidic conditions.

Carabid Ground beetles.

Community A group of species living in the same habitat.

Corrie Corries are steep sided hollows with gently sloping floors that are found in mountainous regions which have experienced glacial activity. (Coire in Gaelic and Cwm in Welsh.)

Cyclonic At temperate latitudes a low-pressure region is called a cyclonic region or a depression. A high-pressure region is known as an anticyclone. Cyclonic weather systems are associated with unsettled, wet weather and strong to gale force winds. Winds circulate in an anticlockwise direction.

Ectotherm An organism whose temperature is determined largely by external conditions.

Eutrophic Water that is rich in organic or mineral nutrients often derived from agricultural or domestic processes.

Headdyke wall or fence marking the upper limit of enclosed fields on a farm.

Home range Many animals spend most of their time within a confined area, their home range, which encompasses foraging range and nest sites. Within this home range a smaller area known as the territory may be defended against other individuals.

Ichneumids Ichneumids are a group of solitary parasitoid wasp belonging to the family Ichneumonidae (more than 1200 British species). They do not have a sting. The ovipositor is used to lay eggs inside invertebrate hosts. The ichneumid larvae consume the host, eating its body fluids, fats and eventually its vital organs.

Inbye Also known as in-by. Small enclosed fields that are close to the farmhouse. Traditionally these fields are fertilized.

John Muir Trust A British charity that is devoted to the preservation of wild land. The name is derived from the great American conservationist John Muir who emigrated to America, from Scotland, as a child in the 1840s. The charity owns several important pieces of wild land that are managed in conjunction with local people. (JMT, PO Box 117, Edinburgh EH7 4AD.)

Leaching The removal of soil nutrients in solution from the soil by percolating water. It is most intense in regions with high rainfall and acidic, free-draining soils.

Meristem The part of plant that contains cells that are capable of cell division.

Mesolithic The period between *c.*14000 and *c.*5000 BP that was characterized by the occurrence of small flint tools called microliths.

Metamorphic rocks Rocks that have been derived from other rocks by the action of intense heat or pressure.

Mycorrhiza An association between a fungus and a flowering plant. The fungus lives in close symbiotic association with the roots. The plant can benefit in a number of ways, including more efficient nutrient uptake.

Neolithic The period from *c.*6000 BP to *c.*4500 BP that is characterized by early farming methods and stone and flint tools.

North Atlantic Drift A warm ocean current that flows from the Gulf of Mexico towards NW Europe as a northeastern extension of the Gulf Stream. It has very significant effects on the climate of NW Europe.

Oligotrophic Oxygenated water that has a low nutrient content.

Ombrotrophic Nutrients are supplied from rain water rather than from the substrate.

Peat Partially decomposed organic matter. Full decomposition is prevented by a shortage of oxygen that is a result of waterlogging.

Phenology The timing of events such as flowering.

Pollen analysis Pollen grains can be preserved in peat and some lake sediments. It is possible to recognize some species or genera from patterns on the surface of the pollen grain. Consequently the relative amounts of pollen from different species, at different depths in the peat or lake sediment can be used to reconstruct the vegetation. The prevailing climate can then be inferred from the vegetation.

Population A group of individuals of the same species that have the potential to breed with each other.

Raptor A bird of prey.

RSPB The Royal Society for the Protection of Birds. They own several significant pieces of upland Britain that are managed for their birds.

Scrub All tree or shrub growth, excluding heathers and prostrate shrubs, that is less than 5 m high.

Species-Area relationship The number of species on an 'island' is related to its area such that a ten-fold increase in area is associated with an approximate doubling in the number of species. The relationship is formalized by functions such as $Species = c.Area^z$, where c is a constant and z is the rate of change in the number of species with increasing area.

Standing crop The dry weight of material harvested from a plot. It is also known as the biomass or yield.

Tetraploid A polyploid condition in which each cell has four haploid sets of chromosomes. Polyploidy is very rare in animals but common in plants.

Thermophile An organism that is adapted to living in a warm environment.

Topography The surface features of a region of land.

Xeromorph A plant that has features which help it to reduce water loss.

WWF World Wide Fund for Nature – an international charity devoted to conservation.

SPECIES LIST

•

This section lists the species mentioned in the book, drawing them together into their appropriate taxonomic groupings. Organisms (except viruses) are first divided into one of five kingdoms (Table 6.1). There are several more subdivisions: the major levels used here are in Table 6.2. The

Table 6.1 Major taxonomic groupings used in the species lists

Kingdom	*Organisms*	*Characteristics*
Prokaryotae (Monera)	Bacteria and cyanobacteria	Single-celled, prokaryotic (lack a membrane-bounded nucleus)
Protoctista	Nucleated algae (including seaweeds), protozoa and slime moulds	Those eukaryotes (i.e. possess a membrane-bounded nucleus) that are not fungi, plants or animals. Often single celled, mainly aquatic (including damp environments and the tissues of other species), often autotrophic
Fungi	Fungi and lichens	Eukaryotic, mainly multicellular, develop directly from spores with no embryological development, heterotrophic and often saprophytic
Plantae	Plants including mosses, liverworts, ferns, conifers and flowering plants	Eukaryotic, multicellular, develop directly from an embryo (multicellular young organisms supported by maternal tissue), usually photoautrophic
Animalia	Invertebrate and vertebrate animals	Eukaryotic, multicellular, develop directly from a blastula (hollow ball of cells), heterotrophic

Source: Margulis and Schwartz 1988

Table 6.2 Summary of the five kingdoms

KINGDOM
PHYLUM
CLASS
    Order
        Family
            Species Authority [Common Name]

final division is into species. The naming of organisms in a standard way is important so that you can be certain which species is under consideration. The naming of species follows international conventions.

Every species is identified by a unique scientific name (often incorrectly called the Latin name) consisting of a binomial term (two words, the first is the genus and the second is the species). Sometimes, especially where there are several species that are difficult to separate, a different term (e.g. agg. or sect.) is placed after the genus, to indicate that several species are involved. In other cases, the name may include an x indicating that the organisms (usually a plant) is a cross between two different species. A third term may be used to designate a sub-species or a variety of the species indicated by the binomial term. Note that species and sub-species names are latinized and should be shown in italics with the generic name beginning with a capital letter. The species name is followed by the name of the person who originally described it (the authority). Where the authority is well known (e.g. if they have named many species), an abbreviation may be used (e.g. L. for Linnaeus, F. for Fabicius). An authority in parentheses indicates that the original species name has been altered (e.g. if the organism has been placed in a different genus). In the absence of parentheses around the authority, the original name given is still in use today.

The names used here follow a variety of sources depending on the taxonomic group including Stace (1997) for plants; Corbet and Harris (1991) for mammals; Cramp (1977–94) for birds.

FUNGI
MYCOPHYCOPHYTA (Lichens)
Cladoniaceae
Cladonia arbuscula (Wallr.) Rabenh.
PLANTAE (plants)
BRYOPHYTA (Mosses and Liverworts)
MUSCI (Mosses)
Ptilidiaceae
Anthelia julacea (L.) Dum.
Amblystegiaceae
Cratoneuron commutatum (Hedw.) Roth
Dicranaceae
Dicranum major Sm.
Dicranum starkei Web. & Mohr
Kiaeria starkei (Web. & Mohr) Hagen
Hypnaceae
Hylocomium splendens (Hedw.) B., S. & G.
Rhytidiadelphus loreus (Hedw.) Warnst
Bartramiaceae
Philonotis fontana (Hedw.) Brid.
Polytrichaceae
Polytrichum alpinum Hedw.
Sphagnaceae
Sphagnum auriculatum Schimp.
Sphagnum capillifolium (Ehrh.)
Sphagnum cuspidatum Hoffm.
Sphagnum recurvum P. Beauv.
Sphagnum russowii Warnst.
Sphagnum squarrosum Crome
Sphagnum warnstorfii Roll
Grimmilaes
Racomitrium lanuginosum (Hedw.) Brid. [Woolly fringe moss]
Bryales
Pohlia wahlenbergii var. *glaciallis* (Schliech. Ex Brid.) E. F. Warb.
SPHENOPHYTA
EQUISETOPSIDA (Horsetails)
FILICINOPHYTA (Ferns)
LYCOPSIDA
Lycopodiaceae
Lycopodium annotinum [Interrupted Clubmoss]
PTEROPSIDA
Dennstaedtiaceae (Bracken family)
Pteridium aquilinum (L.) Kuhn [Bracken]
Woodsiaceae (Lady-fern family)
Woodsia alpina (Bolton) S.F. Gray [Alpine Woodsia]
Woodsia ilvensis (L.) R. Br. [Oblong Woodsia]
CONIFEROPHYTA (Conifers)
PINOPSIDA
Pinaceae (Pine family)

Picea sitchensis (Bong.) Carriere [Sitka Spruce]
Pinus sylvestris L. [Scots Pine]
Cupressaceae (Juniper family)
Juniperus communis L. [Common Juniper]
Juniperus communis alpina [Dwarf Juniper]
ANGIOSPERMOPHYTA (Flowering Plants)
MAGNOLIOPSIDA (Dicotyledons)
Myricales
Myricaceae (Bog-myrtle family)
Myrica gale L. [Bog-myrtle]
Fagales
Fagaceae (Beech family)
Fagus sylvatica L. [Beech]
Quercus petraea (Matt.) [Sessile Oak]
Quercus robur L. [Pedunculate Oak]
Quercus rubra L. [Red Oak]
Betulaceae (Birch family)
Betula nana L. [Dwarf Birch]
Betula pendula Roth [Silver Birch]
Betula pubescens Ehrh. [Downy Birch]
Alnus glutinosa (L.) [Alder]
Corylus avellana L. [Hazel]
Caryophyllales
Chenopodiaceae (Goosefoot family)
Koenigia islandica L. [Iceland Purslane]
Caryophyllaceae (Pink family)
Arenaria norvegica Gunnerus ssp. *norvegica* [Arctic Sandwort]
Cerastium arcticum Lange ssp. *edmondstonii* [Edmonston's Mouse-ear]
Lychnis alpina L. [Alpine Catchfly]
Minuratia stricta (Sw.) Hiern. [Bog Sandwort]
Minuratia rubella (Wahlenb.) Hiern. [Alpine Sandwort]
Sagina intermedia Fenzl [Snow Pearlwort]
Sagina x normaniana Lagerh. [Scottish Pearlwort]
Silene acualis (L.) Jacq. [Moss Campion]
Malvales
Tiliaceae (Lime family)
Tilia x europaea L. [Lime]
Nepenthales
Droseraceae (Sundew family)
Drosera rotundiflora L. [Sundew]
Violales
Violaceae (Violet family)
Viola rupestris F. W. Schmidt [Rock Violet]
Salicales
Salicaceae (Willow family)
Populus nigra L. [Black-poplar]
Populus tremula L. [Aspen]
Salix herbacea L. [Dwarf Willow]
Salix lapponum L. [Downy Willow]
Salix reticulata (Pall.) Schult. & Schult.f. [Dwarf Willow]
Capperales
Brassicaceae
Arabis alpina L. [Alpine Rockcress]
Cochlearia micacea E.S. Marshall [Scottish Scurvygrass]
Diapensiales
Diapensiaceae
Diapensia lapponica L. [Diapensia]
Ericales
Empetraceae (Crowberry family)
Empetrum nigrum L. [Crowberry]
Empetrum hermaphroditium Hagerup [Crowberry]
Ericaceae (Heather family)
Andromeda polifolia L. [Bog Rosemary]
Arctostaphylos alpinus (L.) Spreng. [Alpine Bearberry]
Arctostaphylos uva-ursi (L.) Spreng. [Bearberry]
Rhododendron ponticum L. [Rhododendron]
Calluna vulgaris (L.) Hull [Heather]
Erica cinerea L. [Bell Heather]
Erica tetralix L. [Cross-leaved Heather]
Vaccinium myrtillus L. [Bilberry]
Vaccinium vitis-idaea L. [Cowberry]
Primulales
Primulaceae (Primrose family)
Anagallis tenella L. [Bog Pimpernel]
Lysimachia nemorum L. [Yellow Pimpernel]
Rosales
Saxifragaceae (Saxifrage family)
Saxifraga aizoides L. [Yellow Mountain Saxifrage]
Saxifraga cernua L. [Drooping Saxifrage]
Saxifraga cespitosa L. [Tufted Saxifrage]
Saxifraga hirulus L. [Marsh Saxifrage]
Saxifraga rivularis L. [Highland Saxifrage]
Saxifraga rosacea Moench [Irish Saxifrage]
Rosaceae (Rose family)
Alchemilla alpina L. [Alpine Lady's Mantle]
Alchemilla glabra L. [Lady's Mantle]
Dryas octopetala L. [Mountain Avens]
Potentilla fruticosa L. [Shrubby Cinquefoil]
Rubus chamaemorus L. [Cloudberry]

Sibbaldia procumbens L. [Least Cinquefoil]
Sorbus aucuparia L. [Rowan]

Fabales

Fabaceae (Pea family)

Astragalus alpinus L. [Apine Milk-vetch]
Oxytropis campestris (L.) DC. [Yellow Oxytropis]
Ulex europaeus L. [Gorse]

Celastrales

Aquifoliaceae (Holly family)

Ilex aquifolium L. [Holly]

Euphorbiales

Euphorbiaceae (Spurge family)

Mercurialis perennis L. [Dog's Mercury]

Polygalales

Polygalaceae (Milkwort family)

Polygala amara L. Dwarf [Milkwort]

Geraniales

Oxalidaceae (Wood-sorrel family)

Oxalis acetosella L. [Wood-sorrel]

Gentianales

Gentianaceae (Gentian family)

Gentiana nivalis L. [Alpine Gentian]
Gentiana verna L. [Spring Gentian]

Lamiales

Lamiaceae (Deadnettle family)

Thymus praecox Opiz [Thyme]

Scrophulariales

Scrophulariaceae (Figwort family)

Bartsia alpina L. [Alpine Bartsia]
Veronica fruiticans L. [Rock Speedwell]

Lentibulariaceae (Bladderwort family)

Pinguicula vulgaris L. [Butterwort]
Utricularia vulgaris L. [Bladderwort]

Rubiales

Rubiaceae (Bedstraw family)

Galium palustre L. [Marsh Bedstraw]
Galium saxatile L. [Heath Bedstraw]

Asterales

Asteraceae (Daisy family)

Artemisia norvegica Fr. [Norwegian Mugwort]
Cicerbita alpina (L.) Wallr. [Alpine Sow-thistle]
Crepis paludosa (L.) Moench. [Marsh Hawk's-beard]
Erigeron borealis (Vierh.) Simmons [Alpine Fleabane]
Gnaphalium norvegicum Gunnerus [Norwegian Cudweed]
Homogyne alpina (L.) Cass. [Purple Coltsfoot]

LILIIDAE (Monocotyledons)

Juncales

Juncaceae (Rush family)

Juncus acutiflorus L. [Sharp-flowered Rush]
Juncus effusus L. [Soft Rush]
Juncus squarrosus L. [Heath Rush]
Juncus trifidus L. [Three-leaved Rush]

Cyperales

Cyperaceae (Sedge family)

Eriophorum angustifolium Honck. [Common Cottongrass]
Eriophorum vaginatum L. [Hare's Tail-grass]
Carex atrofusca Schkuhr [Small Jet Sedge]
Carex bigelowii Torr. Ex Schwein. [Stiff Sedge]
Carex curta Gooden. [Pale Sedge]
Carex demissa Hornem. [Common Yellow Sedge]
Carex diocia L. [Diocecious Sedge]
Carex echinata Murr. [Star Sedge]
Carex flacca (L.) Reich. [Glaucous Sedge]
Carex lachenalii Schkuhr [Hare's-foot Sedge]
Carex microglochin Wahlenb. [Bristle Sedge]
Carex nigra (L.) Reich. [Common Sedge]
Carex rariflora (Wahlenb.) Sm. [Mountain Bog Sedge]
Carex rostrata Stokes [Bottle Sedge]
Scirpus cespitosus L. [Deer-grass also known as *Trichophorum*]

Poaceae (Grass family)

Deschampsia flexuosa (L.) Trin. [Wavy Hair grass]
Festuca ovina L. [Sheep's Fescue]
Festuca rubra L. agg. [Red Fescue]
Festuca vivipara (L.) Sm. [Viviparous Fescue]
Molinia caerulea (L.) Moench. [Purple Moor grass]
Nardus stricta L. [Moor Mat grass]
Poa flexuosa Sm. [Wavy Meadow grass]
Agrostis canina L. [Bent Grass]
Agrostis capillaris L. [Common Bent]

Liliales

Liliaceae (Lily family)

Colchicum autumnale L. [Autumn Crocus]
Lloydia serotina (L.) Rohb. [Snowdon Lily]

Polygonatum multiflorum (L.) All. [Solomon's Seal]
Orchidales
Orchidaceae (Orchid family)
Listera cordata (L.) R. Br. [Lesser Twayblade]
ANIMALIA (animals)
PLATYHELMINTHES (flat worms)
Tricladida
Turbellaria
Crenobia alpina
NEMATODA (round worms)
Strongylida
Trichostrongyloidea
Trichostrongylus tenuis Mehlis
ARTHROPODA (arthropods)
INSECTA
Odonata (dragonflies and damselflies)
Coenagrionidae
Cordulegaster boltoni (Donovan)
Plecoptera (stoneflies)
Nemouridae
Protonemura montana Kimmins
Ephemoptera (mayflies)
Siphonuridae
Ameletus inopinatus (Eaton)
Coeloptera (Beetles)
Coccinellidae (ladybirds)
Adalia bipunctata (L.) [two-spot ladybird]
Coccinella hieroglyphica L. [ladybird]
Chrysomelidae (leaf beetles)
Chrysolina cerealis L. [Snowdon leaf beetle]
Plateumaris discolor Panzer [reed beetle]
Lepidoptera (butterflies and moths)
Geometridae
Biston betularia (L.) [peppered moth]
Noctuidae
Celena haworthii Curtis [Haworth's minor]
Cerapterix graminis L. [antler moth]
Coleophoridae
Coleophora alticollella Zeller [rush moth]
Lasiocampidae
Macrothylacia rubi L. [fox moth]
Nymphalidae
Mellicta athalia Rottemberg [heath fritillary]
Argynnis adippe Denis and Schiffermüller [high brown fritillary]
CHORDATA
OSTEICHTHYES (bony fish)
Salmonidae
Coregonus albula L. [vendace]
Salmo salar L. [salmon]
Salmo trutta L. [brown trout]
Salvelinus alpinus L. [charr]
AMPHIBIA (amphibians)
Ranidae (riparian frogs)
Rana temporaria L. [common frog]
REPTILIA (reptiles)
Lacertidae
Lacerta vivipara L. [common lizard]
Viperidae
Vipera berus L. [adder]
AVES (birds)
Anseriformes (swans, geese and ducks)
Aythyinae (bay and sea ducks)
Melanitta nigra L. [common scoter]
Gaviiformes (divers)
Gaviidae
Gavia stellata Pontoppidan [red-throated diver]
Gavia artica L. [black-throated diver]
Falconiformes (birds of prey)
Accipitridae (kites, vultures, harriers, hawks, buzzards and eagles)
Aquila chrysaetos L. [golden eagle]
Haliaeetus albicilla [white-tailed eagle]
Circus cyaneus L. [hen harrier]
Falconidae (falcons)
Falco tinnunculus L. [kestrel]
Falco columbarius L. [merlin]
Falco peregrinus Tunstall [peregrine falcon]
Galliformes (grouse and pheasants)
Tetranoidae (grouse)
Lagopus lagopus scoticus (Lath.) [red grouse]
Lagopus mutus Montin [ptarmigan]
Tetrao urogallus L. [capercaille]
Tetrao tetrix L. [black grouse]
Charadiiformes (shorebirds, gulls and auks)
Charadriidae (plovers)
Charadrius morinellus L. [dotterel]
Pluvialis apricaria L. [golden plover]
Vanellus vanellus L. [lapwing]
Scolopacidae (sandpipers, stints, godwits, curlews, snipe and phalaropes)
Actitis hypoleucos L. [common sandpiper]
Calidris alpina L. [dunlin]
Calidris maritima Br. [purple sandpiper]
Calidris temminckii Leisler [temminck's stint]
Numenius arquata L. [curlew]

Numenius phaeopus L. [whimbrel]
Tringa glareola L. [wood sandpiper]
Tringa nebularia Gunnerus [greenshank]
Stercorariidae (skuas)
Stercorarius parasiticus L. [arctic skua]
Stercorarius skua Brünnich [great skua]
Strigiformes (owls)
Strigidae
Asio flammeus Pontoppidan [short-eared owl]
Nyctea scandiaca L. [snowy owl]
Passeriformes
Cinclidae (dippers)
Cinclus cinclus L. [dipper]
Motacillidae (pipits and wagtails)
Anthus pratensis L. [meadow pipit]
Motacilla cinerea Tunstall [grey wagtail]
Turdidae (robins, chats and thrushes)
Turdus torquatus L. [ring ouzel]
Corvidae (crows)
Corvus corax L. [raven]
Fringillidae (finches)
Acanthis flavirostris L. [twite]
Emberizidae (buntings)
Plectrophenax nivalis [snow bunting]
MAMMALIA (mammals)
Insectivora (insectivores)
Soricida
Sorex araneus L. [common shrew]
Talpa europea L. [mole]
Lagomorpha
Leporidae (rabbits and hares)
Oryctolagus cuniculus (L.) [rabbit]
Lepus capensis Petter [brown hare]
Lepus timidus L. [mountain hare]
Ochotonidae (pikas)
Ochotona pusilla Pallas [steppe pika]
Rodentia (rodents)
Castoridae (beavers)
Castor fiber L. [european beaver]
Muridae (voles, rats and mice)
Clethrionomys glareolus (Schreber) [bank vole]
Microtus agrestis (L.) [field vole]
Microtus oeconomus Pallas [northern vole]
Apodemus sylvaticus (L.) [wood mouse]
Carnivora (terrestrial carnivores)
Canidae (dogs)
Vulpes vulpes (L.) [fox]
Canis lupus L. [wolf]
Mustelidae
Mustela erminea L. [stoat]
Martes martes L. [pine marten]
Ursidae (bears)
Ursus acrtos L. [brown bear]
Felidae (cats)
Felis silvestris Schreber [wild cat]
Artiodactyla
Cervidae (deer)
Cervus elaphus L. [red deer]
Alces alces L. [elk]
Magaceros giganteus [Irish giant elk]
Bovidae
Capra hircus L [feral goat]
Ovis aries L. [domestic sheep]
Bos primigenius Bojanus [auroch]
Suidae
Sus scrofa L. [wild boar]
Perissodactyla
Equidae (horses)
Equus ferus Boddaert [wild horse]
Proboscidea
Mammuthus primigenius [mammoth]

FURTHER READING

The list below gives some of the major texts that will assist in the study of the habitats discussed in this book. They are listed under sections dealing with practical techniques (including the identification of plants and animals), practical conservation, journals and internet resources.

PRACTICAL TECHNIQUES

Several texts help with the design of experiments and methods of approaching a particular ecological problem:

Chalmers, N. and Parker P. (1989) *The Open University project guide.* 2nd edition. Field Studies Occasional Publications No. 9.

Gilbert, D., Horsfield, D. and Thompson, D. B. A. (1997) *The Ecology and Restoration of Montane and Subalpine Scrub Habitats in Scotland.* Scottish Natural Heritage Review No. 83. Scottish Natural Heritage, Edinburgh.

Gilbertson, D. D., Kent, M. and Pyatt, F. B. (1985) *Practical Ecology for Geography and Biology.* Unwin Hyman, London.

Goldspink, C. R., King, S. and Putman, R. J. (1998) *Population Ecology, Management and Welfare of Deer.* Manchester Metropolitan University, Manchester.

Jerram, R. and Drewitt, A. (1998) Assessing Vegetation Condition in the English Uplands. English Nature Research Reports No. 264. English Nature, Peterborough.

Williams, G. (1987) *Techniques and Field Work in Ecology.* Bell and Hyman, London.

Data Analysis

It is important to incorporate statistical techniques into the experimental design, since a poorly designed experiment can be difficult to interpret. There are a number of reference texts, but most are quite heavy going. The following are a few of the more user-friendly student texts:

Chalmers, N. and Parker P. (1989) *The Open University Project Guide.* 2nd edition. Field Studies Occasional Publications No. 9.

Ebden, D. (1987) *Statistics in Geography.* Blackwell, Oxford.

Fowler, J. and Cohen, L. (1990) *Practical Statistics for Field Biology.* Open University Press, Milton Keynes.

Watt, T. A. (1993) *Introductory Statistics for Biology Students.* Chapman & Hall, London.

Identification

There are two types of identification guide. Some are descriptive and usually contain colour illustrations. Care should be taken when using descriptive guides since it is easy to confuse superficially similar species. Better are those texts that incorporate keys (where organisms are sequentially separated out using diagnostic characters). The following section describes some texts which may be of use.

Field guides are accessible descriptive guides to either specific plants or animals, or to particular habitats. Collins and Countrylife publish mainly descriptive guides, while Warne keys to wildflowers, birds and trees provide an alternative which contain identification keys. Field Studies Council AIDGAP keys are user-friendly and cover a wide range of plant and animal groups, especially invertebrates. Naturalists' Handbooks (Richmond Publishing Company) are user-friendly keys to either specific groups of invertebrate animals (mainly insects), or to the occupants of particular habitats. Other identification texts exist which are not part of a series (e.g. Skinner 1984; Marshall and Haes 1988).

More specialist keys are aimed towards the professional and may be difficult for the beginner. However, they are usually more complete than the examples given above. These include the keys produced by the Freshwater Biological Association for the identification of British freshwater invertebrates, the Linnean Society Synopses of the British Fauna covering a large number of invertebrate groups (e.g. earthworms, harvestmen, woodlice, millipedes), and the Royal Entomological Society of London handbooks for the identification of British insects. Other texts such as Stace (1997) do not form part of a series.

PRACTICAL CONSERVATION

Baines, C. and Smart, J. (1984) *A Guide to Habitat Creation.* Ecology Handbook No. 2, Greater London Council, London.
Parker, D. M. (1995) *Habitat Creation – a Critical Guide.* English Nature Science No. 21. English Nature, Peterborough.
Tait, J., Lane, A. and Carr, S. (1988) *Practical Conservation: Site Assessment and Management Planning.* Hodder & Stoughton, Kent.

The British Trust for Conservation Volunteers produce a number of practical conservation handbooks covering woodlands, hedges, paths, drystone walls and wetlands.

JOURNALS

Several journals cover ecology, management and conservation issues including those related to upland areas, for example:

Biological Conservation;
British Wildlife;
Environmental Management;
Journal of Applied Ecology;
Journal of Ecology;
Bird Study;
Scottish Birds.

INTERNET RESOURCES

The following short list of internet addresses should provide a reasonable introduction to pages that are relevant to the British uplands. Please note that Internet addresses may have changed since publication. Try using a search engine to find related resources.

Searching the Flora Europaea
URL=http://www.rbge.org.uk/forms/fe.html

Finding Natural History Books
URL=http://www.nhbs.co.uk/booknet/booknet.html

John Muir Trust
URL=http://www.ma.hw.ac.uk/jmt/

Scottish Natural Heritage (NW Division)
URL=http://www.uhi.ac.uk/snh/index.htm

Lake District National Park
URL=http://www.lake-district.gov.uk/

Dartmoor National Park
URL=http://www.dartmoor-npa.gov.uk/dnp/factfile/homepage.html

The Pennine Way
URL=http://www.pennineway.demon.co.uk/index.htm

The Heather Trust
URL=http://www.electricscotland.com/heather/research.htm

Taxonomy (The Tree of Life)
URL=http://phylogeny.arizona.edu/tree/phylogeny.html

English Nature
URL=http://www.english-nature.org.uk/

REFERENCES

Allan, D. S. (1995) Habitat selection by Whinchats: a case for bracken in the uplands. Pp. 200–205 in D. B. A. Thompson, A. J. Hester and M. B. Usher (eds) *Heaths and Moorland: Cultural Landscapes*. HMSO, London.

Anderson, P. and Radford, E. (1994) Changes in vegetation following reduction in grazing pressure on the National Trust's Kinder Estate, Peak District, Derbyshire, England. *Biological Conservation*, 69: 55–63.

Anon. (1995) The habitats directive – selecting the UK sites. *British Wildlife*, 6: 297–306.

Ashmole, N. P., Nelson, J. M., Shaw, M. R. and Garside, A. (1983) Insects and spiders on snowfields in the Cairngorms, Scotland. *Journal of Natural History*, 17: 599–613.

Baddeley, J. A., Thompson, D. B. A. and Lee, J. A. (1994) Regional and historical variation in the nitrogen content of *Racomitrium lanuginosum* in Britain in relation to atmospheric nitrogen deposition. *Environmental Pollution*, 84: 189–196.

Baines, D. (1996) The implications of grazing and predator management on the habitats and breeding success of black grouse *Tetrao tetrix*. *Journal of Applied Ecology*, 33: 54–62.

Baines, D. and Summers, R. W. (1997) Assessment of bird collisions with deer fences in Scottish forests. *Journal of Applied Ecology*, 34: 941–948.

Barrington, R. M. (1915) The last (?) Irish Golden Eagle. *Irish Naturalist*, 24: 63.

Bartley, D. (1967) Vegetation In M. W. Beresford and G. R. J. Jones (eds) *Leeds and its Regions*, British Association for the Advancement of Science, London.

Bartley, D. (1975) Pollen analytical evidence for Prehistoric forest clearance in the upland area West of Rishworth, W. Yorkshire. *The New Phytologist*, 74: 375–381.

Battarbee, R. W., Jones, V. J., Flower, R. J., Appleby, P. G., Rose, N. L. and Rippey, B. (1996) Palaeolimnological evidence for the atmospheric contamination and acidification of high Cairngorm lochs, with special reference to Lochnagar. *Botanical Journal of Scotland*, 48: 79–87.

Baxter, E. V. and Rintoul, L. J. (1953) *The Birds of Scotland*. Oliver and Boyd, London.

Bayfield, N. G. (1974) Burial of vegetation by erosion debris near chairlifts on Cairngorm. *Biological Conservation*, 6: 246–251.

Bayfield, N. G., Watson. A. and Miller G. R. (1988) Assessing and managing the effects of recreational use on British hills. Pp. 399–414 in M. B. Usher and D.B.A. Thompson (eds) *Ecological Change in the Uplands*. Blackwell Scientific Publications, Oxford.

Bazely, D. R., Vicari, M., Emmerich, S., Filip, L., Lin, D. and Inman, A. (1997) Interaction between herbivores and endophyte-infected *Festuca rubra* from the Scottish islands of St. Kilda, Benbecula and Rum. *Journal of Applied Ecology*, 34: 847–860.

Beecham, J. J. and Kochert M. N. (1975) Breeding biology of the Golden Eagle in southwestern Idaho. *Wilson Bulletin*, 87: 506–513.

Berry, R. J. (1977) *Inheritance and Natural History*. Collins, London.

Bibby, C. J. and Etheridge, B. (1993) Status of the hen harrier *Circus-cyaneus* in Scotland in 1988–89. *Bird Study*, 40: 1–11.

Bibby, C. J., Burgess, N. D. and Hill, D. A. (1993) Status of the hen harrier *Circus cyaneus* in Scotland 1988–89. *Bird Study*, 40: 1–11.

Birks, H. J. B. (1988) Long term ecological change in British uplands. Pp. 37–56 in M. B. Usher and D. B. A. Thompson (eds) *Ecological Change in the Uplands*. Blackwell Scientific Publications, Oxford.

Blaxter, K. L., Kay, R. N. B., Sharman, G. A. M., Cunningham, J. M. M. and Hamilton, W. J. (1974) *Farming the Red Deer*. HMSO, Edinburgh.

Bowman, J. E. (1986) *The Highlands and Islands. A Nineteenth Century Tour*. Alan Sutton, Gloucester.

Brown, A. (1992) *The UK Environment*. The Department of the Environment, Government Statistical Service, HMSO, London.

Brown, A. F. and Bainbridge, I. P. (1995) Grouse moors and upland breeding birds. Pp. 51–66 in D. B. A. Thompson, A. J. Hester and M. B. Usher (eds) *Heaths and Moorland: Cultural Landscapes*, HMSO, London.

Brown, L. H. and Watson, A. (1964) The Golden Eagle in relation to its food supply. *Ibis*, 106: 78–100.

Brown, R. (1799) *General View of the Agriculture of the West Riding*. Board of Agriculture. Halifax.

Bryson, T. (1995) The idea of wilderness. *The John Muir Trust Journal and News*, 19: 25–28.

Budd, J. T. C. (1991) Remote sensing techniques for monitoring land-cover. Pp. 33–59 in F. B. Goldsmith (ed.) *Monitoring for Conservation and Ecology*. Chapman & Hall, London.

Bunce, R. G. H. and Barr, C. J. (1988) The extent of land under different management regimes in the uplands and the potential for change. Pp. 415–426 in M. B. Usher and D. B. A. Thompson (eds) *Ecological Change in the Uplands*. Blackwell Scientific Publications, Oxford.

Bunce, R. G. H., Barr, C. J., Clarke, R. T. and Howard, D. C. (1996) Land classification for strategic ecological survey. *Journal of Environmental Management*, 47: 37–60.

Butterfield, J., Luff, M. L., Baines, M. and Eyre, M. D. (1995) Carabid beetle communities as indicators of conservation potential in upland forests. *Forest Ecology and Management*, 79: 63–77.

Cairngorms Working Party (1993) *Common Sense and Sustainability: A Partnership for the Cairngorms*. The Scottish Office, Edinburgh.

Callander, R. F. and Mackenzie, N. M. (1991) *The Management of Wild Red Deer in Scotland*. Rural Forum Scotland.

Cameron, A. G. (1923) *The Wild Red Deer of Scotland*. Blackwood and Sons, Edinburgh and London.

Chambers, F. M., Dresser, P. Q. and Smith, A. G. (1979) Radiocarbon dating evidence on the impact of atmospheric pollution on upland peats. *Nature*, 282: 829–823.

Clutton-Brock, T. H. and Albon, S. D. (1989) *Red Deer in the Highlands*. BSP Professional Books, Oxford.

Clutton-Brock, T. H. and Albon, S. D. (1992) Trial and error in the Highlands. *Nature*, 358: 11–12.

Collopy, M. W. (1983) A comparison of direct observations and collections of prey remains in determining the diet of Golden Eagles. *Journal of Wildlife Management*, 47: 360–368.

Conrad, V. (1946) Usual formulas of continentality and their limits of validity. *Transactions of the American Geophysics Union*, 27: 663–664.

Coope, G. R. (1977) Fossil Coleopteran assemblages as sensitive indicators of climatic changes during the Devensian (last) cold stage. *Philosophical Transactions of the Royal Society of London*, B280: 313–40.

Corbet, G. B. and Harris, S. (eds) (1991) *The Handbook of British Mammals*. Third Edition. Blackwell Science, Oxford.

Coulson, J. C. (1978) Terrestrial animals. Pp. 160–177 in A. R. Clapham (ed.) *Upper Teesdale: The Area and Its Natural History. Collins, London.*

Coulson, J. C. (1988) The structure and importance of invertebrate communities on peatlands and moorlands, and effects of environmental and management changes. Pp. 365–380 in M. B. Usher and D. B. A. Thompson (eds) *Ecological Change in the Uplands*. Blackwell Scientific Publications, Oxford.

Cramp, S. (1977–94) *Handbook of the birds of Europe, the Middle East and North Africa; the Birds of the Western Palearctic*, Volumes 1, 2, 3, 4, 5 and 8. Oxford University Press, Oxford.

Crick, H. A. P. and Ratcliffe, D. A. (1995) The peregrine, *Falco peregrinus*, breeding population of the United Kingdom in 1991. *Bird Study*, 42: 1–19.

Cross, R. N. R. (1997) Otterburn: the land between. *Sanctuary*, 27: 24–27.

Crump, W. B. (1938) The Little Hill Farm. Halifax Antiquarian Society Papers. (no volume number) 115–196.

Dennis, R. H., Ellis, P. M., Broad, R. A. and Langslow, D. R. (1984) The status of the Golden Eagle in Britain. *British Birds*, 77: 592–607.

English Nature (1995) Establishing criteria for identifying critical natural capital in the terrestrial environment. *English Nature Research Report* No. 141. English Nature, Peterborough.

Erikstad, K. E. (1985) Growth and survival of willow grouse chicks in relation to home range size, brood movements and habitat selection. *Ornis Scandinavica*, 16: 181–190.

Etheridge, B., Summers, R. W. and Green, R. (1997) The effects of human persecution on the population dynamics of hen harriers *Circus cyaneus* nesting in Scotland. *Journal of Applied Ecology*, 34: 1081–1106.

Evans, W. C. (1986) The acute diseases of animals. Pp. 121–132 in Smith, R. T. and Taylor, J. A. (eds) *Bracken: Ecology, Land Use and Control Technology*, Parthenon Press, Carnforth, 121–132.

Everett, M. J. (1971) The Golden Eagle survey in Scotland in 1964–68. *British Birds*, 64: 49–56.

Faull, M. L. and Moorhouse, S. A. (1981) *West Yorkshire: An Archaeological Survey to* AD 1500. West Yorkshire County Council, Wakefield.

Forestry Authority (1998) *The UK Forestry Standard. The Government's Approach to Sustainable Forestry*. Forestry Commission, Edinburgh.

French, D. D., Miller, G. R. and Cummins, R. P. (1997) Recent development of high altitude *Pinus sylvestris* scrub in the northern Cairngorm mountains, Scotland. *Biological Conservation*, 79: 133–144.

Fryday, A. M. (1996) The lichen vegetation of some previously overlooked high level habitats in North Wales. *Lichenologist*, 28: 521–541.

Fryer, G. (1987) Evidence for the former breeding of the Golden Eagle in Yorkshire. *Naturalist*, 112: 3–7

Fuller, R. M., Groom, G. B. and Jones, A. R. (1994) The land cover map of Great Britain: an automated classification of Landsat Thematic Mapper data. *Photogrammetric Engineering and Remote Sensing*, 60: 553–562.

Galpin, O. P., Whitaker, C. J., Whitaker, R. and Kassab, J. Y. (1990) Gastric cancer in Gwynedd. Possible links with bracken. *British Journal of Cancer*, 61: 737–40.

Gardner, S. M., Hartley, S. E., Davies, A. and Palmer, S. C. F. (1997) Carabid communities on heather moorlands in north eastern Scotland: the consequences of grazing pressure for community diversity. *Biological Conservation*, 81: 275–286.

Gilbert, O. L. and Fryday, A. M. (1996) Observations on the lichen flora of high ground in the west of Ireland. *Lichenologist*, 28: 113–127.

Gimingham, C. H. (1992) *The Lowland Heath Management Handbook*. English Nature, Peterborough.

Gliessman, S. R. (1976) Allelopathy in a broad spectrum of environments as illustrated by bracken. *Botanical Journal of the Linnean Society*, 73: 95–104.

Grace, J. and Unsworth, M. H. (1988) Climate and microclimate of the uplands. Pp. 137–150 in M. B. Usher and D. B. A. Thompson (eds) *Ecological Change in the Uplands*. Blackwell Scientific Publications, Oxford.

Grant, S. A. and Armstrong, H. M. (1993) Grazing ecology and the conservation of heather moorland: the development of models as aids to management, *Biodiversity and Management*, 2: 79–94.

Green, R. E. (1996) The status of the Golden Eagle in Britain in 1992. *Bird Study*, 43: 20–27.

Grieve, I. C., Davidson, D. A. and Gordon, J. E. (1995) Nature, extent and severity of soil-erosion in upland Scotland. *Land Degradation and Rehabilitation*, 6: 41–55.

Grubb, P. (1974) The rut and behaviour of Soay rams. Pp. 195–241 in P. A. Jewell, C. Milner and J. Morton-Boyd (eds) *Island Survivors: the Ecology of Soay Sheep of St. Kilda*. Athlone Press, London.

Grubb, P. and Jewell, P. A. (1974) Movement, daily activity and home range of Soay sheep. Pp. 160–194 in P. A. Jewell, C. Milner and J. Morton-Boyd (eds) *Island Survivors: the Ecology of Soay Sheep of St. Kilda*. Athlone Press, London.

Hadfield, P. R. and Dyer, A. F. (1986). Polymorphism of cyanogenesis in British populations of bracken (*Pteridium aquilinum* (L.) Kuhn). Pp. 293–300 in R. T. Smith and J. A. Taylor (eds), *Bracken: Ecology, Land Use and Control Technology*, Parthenon Press, Carnforth.

Harrison, S. J. (1973) *An Ecoclimatic Gradient in North Cardiganshire, West Central Wales*. Unpublished Ph.D. dissertation, University College of Wales, Aberystwyth, Department of Geography.

Haworth, P. F. and Thompson, D. B. A. (1990) Factors associated with the breeding distribution of upland birds in the South Pennines, England. *Journal of Applied Ecology*, 27: 562–577.

Hester, A. J., Miller, D. R. and Towers, W. (1996)

Landscape-scale vegetation change in the Cairngorms, Scotland, 1946–1988: Implications for land management. *Biological Conservation*, 77: 41–51.

Hewson, R. (1991) Mountain hare/Irish hare *Lepus timidus*. Pp. 161–185 in G. B. Corbett and S. Harris (eds) *The Handbook of British Mammals*, Third Edition. Blackwell, Oxford.

Holland. P .K., Robson, J. E. and Yalden, D. W. (1982) The Status and distribution of the common sandpiper in the Peak District. *Naturalist*, 107: 77–86.

Holloway, S. (1996) *The Historical Atlas of Breeding Birds in Britain and Ireland*, 1875–1900. Poyser, London.

Holloway, W. (1967) *The Effects of Red Deer and Other Animals on Naturally Regenerated Scots Pine*. Unpublished Ph.D thesis. University of Aberdeen.

Hope, D., Picozzi, N., Catt, D. C. and Moss, R. (1996) Effects of reducing sheep in the Scottish Highlands. *Journal of Range Management*, 49: 301–310.

Hopkins, A., Wainwright, J., Murray, P. J., Bowling, P. J. and Webb, M. (1988) Survey of upland grassland in England and Wales: changes in age structure and botanical composition since 1970–72 in relation to grassland management and physical features. *Grass and Forage Science*, 43: 185–98.

Hopkins, J. J. (1995) List of possible Special Areas of Conservation in the UK. *British Wildlife*, 6: 286–296.

Horsfield, D. and Thompson, D. B. A. (1996) The uplands: guidance on terminology regarding altitudinal zonation and related terms. *Information and Advisory Note* No. 26, Scottish Natural Heritage, Edinburgh.

Hudson, P. J. (1986a) *Red Grouse: the Biology and Management of a Wild Gamebird*. Fordingbridge: The Game Conservancy Trust.

Hudson, P. J. (1986b) Bracken and ticks on grouse moors in the north of England. Pp. 161–170 in R.T. Smith and J. A Taylor (eds) *Bracken: Ecology, Land Use and Control Technology*, Parthenon Press, Carnforth.

Hudson, P. J. (1992) *Grouse in Space and Time – the Population Biology of a Managed Gamebird*. Fordingbirde: The Game Conservancy Trust.

Hudson, P. J. and Dobson, A. P. (1990) Red grouse population cycles and the population dynamics of the caecal nematode *Trichostrongylus tenuis*. Pp. 5–20 in A. N. Lance and J. H. Lawton (eds) *Red Grouse Population Processes*. Royal Society for the Protection of Birds, Sandy.

Hudson, P. J., Dobson, A. P. and Newborn, D. (1985) Cyclic and non-cyclic populations of red grouse: a role for parasites? Pp. 77–89 in D. Rollinson and R. M. Anderson (eds) *Ecology and Genetics of Host-parasite Interactions*. Academic Press for the Linnean Society, London.

Jacobi, R. M., Tallis, J. H. and Mellars, P. A. (1976) The Southern Pennine Mesolithic and the ecological record. *Journal of Archaeological Science*, 3: 307–320.

Jenkins, D., Watson, A. and Miller G. R. (1967) Population fluctuations in the red grouse *Lagopus lagopus scoticus*. *Journal of Animal Ecology*, 36: 97–122.

Jeremy, A. C. and Tutin, T. G. (1968) *British Sedges*. Botanical Society of the British Isles, London.

JNCC (1995) *Biodiversity: The UK Steering Group Report. Volume 2: Action Plans*. HMSO, London.

Jondottir, I. S. (1991) Effects of grazing on tiller size and population dynamics of a clonal sedge (*Carex bigelowii*). *Oikos*, 62: 177–188.

Jones, A. (1991) British wildlife and the law: a review of the species protection provisions of the Wildlife and Countryside Act 1981. *British Wildlife*, 2: 345–358.

Latusek, E. P. (1983) *The Autecology of* Molinia caerulea *(L) Moench with Particular Reference to Grazing*. Unpublished Ph.D. thesis, Manchester Polytechnic.

Lavers, C. P. and Haines-Young, R. H. (1997) Displacement of dunlin *Calidris alpina* by forestry in the Flow Country and an estimate of the value of moorland adjacent to plantations. *Biological Conservation*, 79: 87–90.

Lawton, J. H. (1976) The structure of the arthropod community on bracken. *Botanical Journal of the Linnean Society*, 73: 187–216.

Lawton, J. H. (1990) Red grouse populations and moorland management. Pp. 84–99 in A. N. Lance and J. H. Lawton (eds) *Red Grouse Population Processes*. Royal Society for the Protection of Birds, Sandy.

Lawton, J. H., Macgarvin, M. and Heads, P. A.

(1986) The ecology of bracken-feeding insects: background for a biological control programme. Pp. 285–92 in R.T. Smith and J. A Taylor (eds) *Bracken: Ecology, Land Use and Control Technology*, Parthenon Press, Carnforth.

Lee, J. A., Tallis, J. H. and Woodin, S. J. (1988) Acidic deposition and British upland vegetation. Pp. 151–162 in M. B. Usher and D. B. A. Thompson (eds) *Ecological Change in the Uplands*. Blackwell Scientific Publications, Oxford.

Lines, R. (1984) Species and seed origin trials in the industrial Pennines. *Quarterly Journal of Forestry*, 78: 9–73.

Lockie, J. D. (1964) The breeding density of the Golden Eagle and Fox in relation to food supply in Wester Ross, Scotland. *Scottish Naturalist*, 71: 67–77.

Lockie, J. D. and Ratcliffe, D. A. (1964) Insecticides and Scottish Golden Eagles. *British Birds*, 57: 89–102.

Lockie, J. D., Ratcliffe, D. A. and Balharry, R. (1969) Breeding success and organochlorine residues in Golden Eagles in West Scotland. *Journal of Applied Ecology*, 6: 381–389.

Love, J. A. (1983) *The Return of the Sea Eagle*. University Press, Cambridge.

Lowe, J. J. and Walker, M. J. C. (1984) *Reconstructing Quaternary Environments*. Longman, Harlow.

Lynch, J. J., Hinch, G. N. and Adams, B. B. (1992) *The Behaviour of Sheep: Biological Principles and Implications for Production*. CAB International.

McClatchley, J. (1996) Spatial and altitudinal gradients of climate in the Cairngorms – observations from climatological and automatic weather stations. *Botanical Journal of Scotland*, 48: 31–49.

MacKay, J. W. (1995) People, perceptions and moorland. Pp. 102–114 in D. B. A. Thompson, A. J. Hester and M. B. Usher (eds) *Heaths and Moorland: Cultural Landscapes*, HMSO, Edinburgh.

Mackenzie, N. A. (1987) *The Native Woodlands of Scotland*. Friends of the Earth (Scotland), Edinburgh.

Maclean, M. and Carrell, C. (1986) *As an Fhearann: From the Land. Mainstream*, Edinburgh.

MacPherson, H. A. (1892) *A Vertebrate Fauna of Lakeland*. David Douglas, Edinburgh.

McVean, D. N. and Ratcliffe, D. A. (1962) Plant communities of the Scottish Highlands. *Nature Conservancy Monograph No 1*. HMSO, London.

Manly, G. (1970) The climate of the British Isles. Pp. 81–133 in C. C. Wallen (ed.) *Climates of Northern and Western Europe*. Elsevier (World Survey of Climatology, Volume 5), Oxford.

Margulis, L. and Schwartz, K. V. (1988) *Five Kingdoms*, Second edition. Freeman, New York.

Marquiss, M., Ratcliffe, D. A. and Roxburgh, R. (1985) The numbers, breeding success and diet of Golden Eagles in southern Scotland in relation to changes in land use. *Biological Conservation*, 33: 1–17.

Marrs, R. H. and Hicks, M. J. (1986) Study of vegetation change at Lakenheath Warren: a re-evaluation of A. S. Watt's theories of bracken dynamics in relation to succession and vegetation management. *Journal of Applied Ecology*, 23: 1029–46.

Marrs, R. H. and Pakeman, R. J. (1995) Bracken invasion – Lessons from the past and prospects for the future. Pp. 200–205 in D. B. A. Thompson, A. J. Hester and M. B. Usher (Eds) *Heaths and Moorland: Cultural Landscape*, HMSO, London.

Marshall, J. A. and Haes, E. C. M. (1988) The grasshoppers and allied insects of the British Isles. *Harley Books*, Essex.

Mason, W. L. and Quine, C. P. (1995) Silvicultural possibilities for increasing structural diversity in British spruce forests – the case of Kielder Forest. *Forest Ecology and Management*, 79: 13–28.

Miles, J. (1988) Vegetation and soil change in the uplands. Pp. 57–70 in M. B. Usher and D. B. A. Thompson (eds) *Ecological Change in the Uplands*. Blackwell Scientific Publications, Oxford.

Miller, D. R. Morrice, J. G. and Whitworth, P. L. (1990) Bracken distribution and spread in upland Scotland: an assessment using digital mapping techniques. Pp. 121–132 in R. T. Smith and J. A Taylor (eds) *Bracken: Ecology, Land Use and Control Technology*, Parthenon Press, Carnforth.

Miller, G. R., Ceddes, C. and Mardon, D. K. (1994) Response of the alpine gentian (*Gentina nivalis)* and other montane species to protection from grazing. *Scottish Natural Heritage Review*, No. 32, Battleby.

Ministry of Agriculture, Fisheries and Food (1976) *The Agricultural Climate of England and Wales. Technical Bulletin 35*, HMSO, London.

Mitchell, B., Staines, B., W. and Welch, D. (1977) *Ecology of Red Deer: a Research Review Relevant to Their Management in Scotland.* Natural Environment Research Council, Institute of Terrestrial Ecology.

Mitchell, F. J. D. and Kirby, K. J. (1990) The impact of large herbivores on the conservation of semi-natural woods in the British uplands. *Forestry*, 63: 333–353.

Moore, P. D. (1973) The influence of prehistoric cultures upon the initiation and spread of blanket bog in upland Wales. *Nature*, 241: 350–353.

Moore, P. D. (1975) Origin of blanket mires. *Nature*, 256: 267–269.

Moorhouse, S. A. (1979) Documentary evidence for the landscape of the Manor of Wakefield during the Middle Ages. *Landscape History*, 1: 44–58.

Moss, C. E. (1900) Changes in the Halifax flora during the last century and a quarter. *The Naturalist*, Hull, 26: 165–171.

Moss, C. E. (1901) Changes in the Halifax Flora During the Last Century and a Quarter. *The Naturalist*, Hull 27: 99–107.

Moss, R. and Watson, A. (1985) Adaptive value of spacing behaviour in population cycles of red grouse and other animals. Pp. 275–284 in R. M. Sibley and R. H. Smith (eds) *Behavioural Ecology. Ecological Consequences of Adaptive Behaviour*. Blackwell Scientific, Oxford.

Moss, R., Watson, A., Rothery, P. and Glennie, N. W. (1981) Clutch size, egg size, hatch weight and laying date in relation to early mortality in red grouse *Lagopus lagopus scoticus* chicks. *Ibis*, 12: 450–462.

Moss, R., Watson, A. and Parr, R. (1996) Experimental prevention of a population cycle in red grouse. *Ecology*, 77: 1512–1530.

Mountford, M. D., Watson, A., Moss, R., Parr, R. and Rothery, P. (1990) Land inheritance and population cycles of red grouse. Pp. 78–83 in A. N. Lance and J. H. Lawton (eds) *Red Grouse Population Processes*. Royal Society for the Protection of Birds, Sandy.

Murray, W. H. (1987) *Scotland's Mountains*. Scottish Mountaineering Club, Glasgow.

Nethersole-Thompson, D. (1973) *The Dotterel*. Collins, London.

Nethersole-Thompson, D. and Watson, A. (1981) *The Cairngorms. Their Natural History and Scenery*. The Melven Press, Perth.

Newton, I. (1979) *Population Ecology of Raptors*. Poyser, Berkhamsted.

Newton, I and Galbraith, C. A. (1991) Organochlorines and mercury in the eggs of Golden Eagles *Aquila chrysaetos* from Scotland. *Ibis*, 133: 115–120.

Nichols, D. (1990) *Safety in Biological Fieldwork – Guidance Notes for Codes of Practice*. The Institute of Biology, London.

Nicholson, I. A. and Paterson, I. S. (1976) The ecological implications of bracken control to plant/animal systems. *Biological Journal of the Linnean Society*, 73: 269–284.

Nowell, J. (1866) Notes on Some Rare Mosses of Todmorden. *The Naturalist*, Hull 3: 1–3.

Omerod, S. J., Rundle, S. D., Lloyd, E. C. and Douglas, A. A. (1993) The influence of riparian management on the habitat structure and macroinvertebrate communities of upland streams draining conifer plantation forests. *Journal of Applied Ecology*, 30: 13–24.

Page, N. (1979) Experimental aspects of fern ecology. Pp. 551–589 in A.F Dyer (ed.). *The Experimental Biology of Ferns*. Academic Press, London.

Pakeman, R. J. and Marrs, R. H. (1992) The conservation value of bracken *Pteridium aquilinum* (L) Kuhn – dominated communities in the UK, and as assessment of the ecological impact of bracken expansion or its removal. *Biological Conservation*, 62: 101–14.

Parr, S. J. (1994) Changes in the population-size and nest sites of merlins *Falco columbarius* in Wales between 1970 and 1991. *Bird Study*, 41: 42–47.

Pearce-Higgins, J. W. and Yalden, D. W. (1997) The effect of resurfacing the Pennine Way on recreational use of blanket bog in the Peak District National Park, England. *Biological Conservation*, 82: 337–343.

Pearsall, W. H. (1971) *Mountains and Moorlands*. Revised edition. Collins (The New Naturalist), London.

Pepin, N. C. (1995) Thermal climate of marginal maritime uplands. *Geografiska Annaler*, 77A: 167–185.

Phillips, J. (1981) Moor burning. Pp. 171–175 in

J. Phillips, D. W. Yalden and J. Tallis (eds) *Peak District Moorland Erosion Study. Phase 1 Report.* Peak Park Planning Board, Bakewell.

Phillips, J. and Watson, A. (1995), Key Requirements for Management of Heather Moorland: Now and For the Future. Pp. 344–361 in D. B. A. Thompson, A. J. Hester and M. B. Usher (eds) *Heaths and Moorland: Cultural Landscapes* HMSO, Edinburgh.

Phillips, J., Watson, A. and Macdonald, A. (1993) *A Muirburn Code.* Scottish Natural Heritage, Battleby.

Phillips, R. L., Wheeler, A. H., Forrester, N. C., Lockhart, J. M. and Mceneaney, T. P. (1990) Nesting ecology of Golden Eagles and other raptors in southeastern Montana and northern Wyoming. *Fish and Wildlife Technical Report No. 26.* USDI, Washington.

Pigott, C. D. (1978) Climate and vegetation. Pp. 102–121 in A. R. Clapham (ed.) *Upper Teesdale: The Area and Its Natural History.* Collins, London.

Potter, C., Barr, C. and Lobley, M. (1996) Environmental change in Britain's countryside: an analysis of recent patterns and socio-economic processes based on the Countryside Survey 1990. *Journal of Environmental Management,* 48: 169–186.

Potts, G. R. (1998) Global dispersion of nesting hen harriers *Circus cyaneus*; implications for grouse moors in the UK. *Ibis,* 140: 76–88

Potts, G. R., Tapper, S. C. and Hudson, P. J. (1984) Population fluctuations in red grouse: analysis of bag records and a simulation model. *Journal of Animal Ecology,* 53: 31–36.

Press, M., Ferguson, P. and Lee, J. (1983) 200 Years of Acid Rain. *The Naturalist,* 108: 125–129.

Preston, C. D. and Hill, M. O. (1997) The geographical relationships of British and Irish vascular plants. *Biological Journal of the Linnean Society,* 124: 1–120.

Ratcliffe, D. A. (1970) Changes attributable to pesticides in egg breakage frequency and eggshell thickness in some British birds. *Journal of Applied Ecology,* 7: 67–107.

Ratcliffe, D. A. (1977) *A Nature Conservation Review: the Selection of Biological Sites of National Importance.* Cambridge University Press (for the Nature Conservancy Council), Cambridge.

Ratcliffe, D. A. (1990a) *Bird Life of Mountain and Upland.* Cambridge University Press, Cambridge.

Ratcliffe, D. A. (1990b) Upland birds and their conservation. *British Wildlife,* 2: 1–12.

Ratcliffe, D. A. (1991) The mountain flora of Britain and Ireland. *British Wildlife,* 3: 10–21.

Ratcliffe, D. A. (1993) *The Peregrine Falcon.* Poyser, Calton.

Ratcliffe, D. A. (1997) *The Raven.* Poyser, London.

Ratcliffe, D. A. and Thompson, D. B. A. (1988) The British uplands: their ecological character and international significance. Pp. 9–36 in M. B. Usher and D. B. A. Thompson (eds) *Ecological Change in the Uplands.* Blackwell Scientific Publications, Oxford.

Ratcliffe, P. R. (1987) *The Management of Red Deer in the Commercial Forests of Scotland Related to Population Dynamics and Habitat Changes.* Unpublished D.Phil. thesis, University of London.

Redpath, S. M. and Thirgood, S. J. (1997) *Birds of Prey and Red Grouse.* The Stationery Office, London.

Redpath, S. M., Madders, M., Donnelly, E., Anderson, B., Thirgood, S. J. and McLeod, D. (1998) Nest site selection by hen harriers in Scotland. *Bird Study,* 45: 51–61.

Rees, R. M. and Ribbens, J. C. H. (1995) Relationships between afforestation, water chemistry and fish stocks in an upland catchment in southwest Scotland. *Water, Air and Soil Pollution,* 85: 303–308.

Rimes, C. A., Farmer, A. M. and Howell, D. (1994) A survey of the threat of surface water acidification to the nature conservation interest of freshwaters on sites of special scientific interest in Britain. *Aquatic Conservation – Marine and Freshwater Ecosystems,* 4: 31–44.

Ritchie, J. C. (1956) Biological flora of the British Isles: *Vaccinium myrtillus L. Journal of Ecology,* 44: 291–199.

Roberts, J., Macdonald, A., and Wood-Gee. V. (1996) Bracken control. Information and advisory note *Scottish Natural Heritage Review,* No. 24, Battleby.

Robinson, R. C. (1986) Practical herbicide use for bracken control. Pp. 231–240 in R. T. Smith and J. A Taylor (eds) *Bracken: Ecology, Land Use and Control Technology*, Parthenon Press, Carnforth.

Rodwell, J. S. (1991a) *British Plant Communities, Volume 2: Mires and Heaths*. Cambridge University Press, Cambridge.

Rodwell, J. S. (1991b) *British Plant Communities, Volume 1: Woodlands and Scrub*. Cambridge University Press, Cambridge.

Rotheray, G. E. and Horsfield, D. (1995) Insects of Scottish mountains. *British Wildlife*, 6: 160–167.

Rowell, T. A. (1990) Management of peatlands for conservation. *British Wildlife*, 1: 144–156.

Salim, K. A., Carter, P. L., Shaw, S. and Smith, C. A. (1988) Leaf abscission zones in *Molinia caerulea* (L) Moench, the purple moor grass. *Annals of Botany*, 62: 429–434.

Scott, M. (1993) A manifesto for the Cairngorms. Pp. 33–40 in A. Watson and J. Conroy (eds) *The Cairngorms: Planning Ahead*. Kincardine and Deeside District Council, Stoneavon.

Scottish Natural Heritage (1994) *Red Deer and the Natural Heritage. SNH Policy Paper*. Scottish Natural Heritage, Battleby.

Scottish Natural Heritage (1995) *The Natural Heritage of Scotland: an Overview*. Scottish Natural Heritage, Edinburgh.

Shaw, G. (1995) Habitat selection by short-eared owls *Asio flammeus* in young coniferous forests. *Bird Study*, 42: 158–164.

Sidaway R. (1990) *Birds and Walkers. A Review of Existing Research on Access to the Countryside and Disturbance to Birds*. Ramblers' Association, London.

Sidaway R. (1994) *Recreation and the Natural Heritage: a Research Review*. Scottish Natural Heritage, Edinburgh.

Skinner, B. (1984) *Moths of the British Isles*. Viking, Middlesex.

Stace, C. (1997) *New Flora of the British Isles*. Second edition. Cambridge University Press, Cambridge.

Staines, B. W. and Ratcliffe, P. R. (1987) Estimating the abundance of red and roe deer and their current status in Great Britain. In S. Harris (ed.) *Mammal Population Studies. Symposium of the Zoological Society of London*, 58: 131–152.

Staines, B. W. and Scott, D. (1994) Recreation and red deer: a preliminary review of the issues. *Scottish Natural Heritage Review*, No. 31, Battleby.

Staines, B. W. and Welch, D. (1989) An appraisal of deer damage in conifer plantations. Pp. 61–76 in R.McIntosh (ed.) *Deer and Forestry*. ICF. Edinburgh.

Steenhof, K, Kochert, M. N. and McDonald, T. L. (1997) Interactive effects of prey and weather on golden eagle reproduction. *Journal of Applied Ecology*, 66: 350–362.

Stevenson, A. C. and Birks, H. J. B. (1995) Heaths and moorland. Long term changes, and interactions with climate and people. Pp. 224–239 in D. B. A. Thompson, A. J. Hester and M. B. Usher (Eds) *Heaths and Moorland: Cultural Landscapes*, HMSO, London.

Stevenson, A. C. and Thompson, D. B. A. (1993) Long-term changes in the extent of heather moorland in upland Britain and Ireland: palaeoecological evidence for the importance of grazing. *The Holocene*, 3: 70–6.

Stewart, F. and Hester, A. (1998) The impact of red deer on woodland and heathland dynamics in Scotland. Pp. 54–60 in Goldspink, C. R., King, S. and Putman, R. J. (eds) *Population Ecology, Management and Welfare of Deer*. Manchester Metropolitan University, Manchester.

Stroud, D. A., Mudge, G. P. and Pienkowski, M. W. (1990) *Protecting Internationally Important Bird Sites*. Nature Conservancy Council, Peterborough.

Tallis, J. H. (1964a) *The Pre-peat Vegetation of the Southern Pennines*. New Phytologist, 63: 363–73.

Tallis, J. H. (1964b) Studies on Southern Pennine Peats. III. The behaviour of *Sphagnum*. *Journal of Ecology*, 52: 345–353.

Tallis, J. H. (1994) Pool and hummock patterning in a Southern Pennine blanket mire II. The formation and erosion of the pool system. *Journal of Ecology*, 82: 789–803.

Tallis, J. H. (1995) Climate and erosion signals in British blanket peats – The significance of *Racomitrum lanuginosum* remains. *Journal of Ecology*, 83: 1021–1030.

Tallis, J. H. (1997) The pollen record of *Empetrum nigrum* in southern Pennine peats: implictions for erosion and climate change. *Journal of Ecology*, 85: 455–465.

Tallis, J. H. and McGuire, J. (1972) Central Rossendale: The evolution of an upland vegetation. I. The clearance of woodland. *Journal of Ecology*, 60: 721–737.

Tallis, J. H. and Livett, E. A. (1994) Pool and

hummock patterning in a southern Pennine blanket mire. I. Stratigraphic profiles for the last 2800 years. *Journal of Ecology*, 82: 775–788.

Taylor, J. A. (1976) Upland climates. Pp. 264–287 in T. J. Chandler and S. Gregory (eds) *The Climate of the British Isles*. Longman, London.

Taylor, J. A. (1986) The bracken problem: a local hazard and global issue. Pp. 21–42 in R. T. Smith and J. A Taylor (eds) *Bracken: Ecology, Land Use and Control Technology*, Parthenon Press, Carnforth.

Taylor, J. A. (1989) Bracken, toxicity and carcinogenicity as related to animal and human health. *Institute of Earth Studies Publication No. 44*, Aberyswyth.

Taylor, K. (1993) Land, wildlife and conservation in the Cairngorms. *British Wildlife*, 5: 152–161.

Thomson, A. G., Radford, G. L., Norris, D. A. and Good, J. E. G. (1993) Factors affecting the distribution and spread of *Rhododendron* in North Wales. *Journal of Environmental Management*, 39: 199–212.

Thompson, D. B. A. and Brown, A. (1992) Biodiversity in montane Britain: habitat variation, vegetation diversity and some objectives for conservation. *Biodiversity and Conservation*, 1: 179–208.

Thompson, D. B. A. and Horsfield, D. (1997) Upland habitat conservation in Scotland: a review of progress and some proposals for action. *Botanical Journal of Scotland*, 49: 501–516.

Thompson, D. B. A. and Whitfield, D. P. (1993) Research on mountain birds and their habitats. *Scottish Birds*, 17: 1–8.

Thompson, D. B. A., Galbraith, H. and Horsfield, D. (1987) Ecology and resources of Britain's mountain plateaux: land use conflicts and impacts. Pp. 22–31 in M. Bell and R. G. H. Bunce (eds) *Agriculture and Conservation in the Hills and Uplands*. Institute of Terrestrial Ecology, Cambridge.

Thompson, D. B. A., Stroud, D. A. and Pienkowski, M. W. (1998) Afforestation and upland birds: consequences for population ecology. Pp. 237–260 in M. B. Usher and D. B. A. Thompson (eds) *Ecological Change in the Uplands*. Blackwell Scientific Publications, Oxford.

Thompson, D. B. A., Horsfield, D., Gordon, J. E. and Brown, A. (1993) The environmental significance of the Cairngorm Massif. Pp. 15–24 in A. Watson and J. Conroy (eds) *The Cairngorms: Planning Ahead*. Kincardine and Deeside District Council, Stoneavon.

Thompson, D. B. A., MacDonald, A. J., Marsden, J. H. and Galbraith, C. A. (1995a) Upland heather moorland in Great Britain: a review of international importance, vegetation change, and some objectives for nature conservation. *Biological Conservation*, 71: 163–178.

Thompson, D. B. A., Hester, A. J. and Usher, M. B. (1995b) *Heaths and Moorlands, Cultural Landscapes*. HMSO, Edinburgh.

Tout, D. (1976) Temperature. Pp. 96–128 in T. J. Chandler and S. Gregory (eds) *The Climate of the British Isles*. Longman, London.

Tudor, G. and Mackey, E. C. (1995) Upland land cover change in post-war Scotland. Pp. 28–42 in D. B. A. Thompson, A. J. Hester and M. B. Usher (eds) *Heaths and Moorland: Cultural Landscapes*, HMSO, Edinburgh.

Usher, M. B. (1992) Management and diversity in *Calluna* heathland. *Biodiversity and Conservation*, 1: 63–79.

Usher, M. B. and Thompson, D. B. A. (1993) Variation in the upland heathlands of Great Britain: Conservation importance. *Biological Conservation*, 66: 69–81.

Ussher, R. J. and Warren, R. (1900) *The Birds of Ireland*. Gurney and Jackson, London.

Veerasekaran, P., Kirkwood, R. C. and Fletcher, W. W. (1976) The mode of action of asulam (methyl(4-aminobenzenesulphonyl) carbamate) in bracken. *Botanical Journal of the Linnean Society*, 73: 247–68.

Veerasekaran, P., Kirkwood, R. C. and Fletcher, W. W. (1977a) Studies on the mode of action of asulam in bracken (*Pteridium aquilinum* (L) Kuhn). I. Absorption and translocation of C asulam. *Weed Research*, 17: 33–9.

Veerasekaran, P., Kirkwood, R. C. and Fletcher, W. W. (1977b) Studies on the mode of action of asulam in bracken (*Pteridium aquilinum* (L) II. Biochemical activity in the rhizome buds. *Weed Research*, 17: 85–92.

Veerasekaran, P., Kirkwood, R. C. and Fletcher, W. W. (1978) Studies on the mode of action of asulam in bracken (*Pteridium acquilinum* (L) Kuhn). III. Long-term contol of field bracken. *Weed Research*, 18: 315–19.

Vincent, P. (1990) *The Biogeography of the British Isles: An Introduction*. Routledge, London.
Wallace, H. L. and Good, J. E. G. (1995) Effects of afforestation on upland plant-communities and implications for vegetation management. *Forest Ecology and Management*, 79: 29–46.
Watson A. (1979) Bird and mammal numbers in relation to human impact at ski lifts in Scottish hills. *Journal of Applied Ecology*, 16: 753–764.
Watson A. (1982) Effects on human impact on ptarmigan and red grouse near ski lifts in Scotland. *Annual Report of the Institute of Terrestrial Ecology*, 1981, 51.
Watson, A. (1983) Eighteenth century deer numbers and pine regeneration near Braemar, Scotland. *Biological Conservation*, 25: 289–305.
Watson A. (1985) Soil erosion and vegetation damage near ski lifts at Cairn Gorm, Scotland. *Biological Conservation*, 33: 363–381.
Watson, A. (1989) Land use, reduction of heather, and natural tree regeneration on open upland. *ITE Annual Report*, HMSO, London.
Watson, A. (1993a) A vision for the Cairngorms and critique of the Working Party's report. Pp. 59–86 in A. Watson and J. Conroy (eds) *The Cairngorms: Planning Ahead*. Kincardine and Deeside District Council, Stoneavon.
Watson, A. (1993b) Defects of fencing for native woodlands. *Native Woodlands Discussion Group Newsletter No 18*, pp. 5355.
Watson, A. (1996a) Internationally important environmental features of the Cairngorms, Research and research needs. *Botanical Journal of Scotland*, 48: 1–12.
Watson, A. (1996b) Human induced increases of Carrion Crows and gulls on Cairngorm plateaux. *Scottish Birds*, 18: 205–213.
Watson, A. (1997) Habitat use by Snow Buntings in Scotland from spring to autumn. *Scottish Birds*, 19: 105–113.
Watson, A. and Rothery, P. (1986) Regularity in spacing of Golden Eagle *Aquila chrysaetos* nests used within years in northeast Scotland. *Ibis*, 128: 406–408.
Watson, A., Payne, S. and Rae, R, (1989) Golden Eagles *Aquila chrysaetos*: land use and food in northeast Scotland. *Ibis*, 131: 336–348.
Watson, J. (1775) *The History and Antiquities of the Parish of Halifax*. Halifax.
Watson, J. (1991) The Golden Eagle and pastoralism across Europe. Pp. 56–57 in D. J. Curtis, E. M. Bignal and M. A. Curtis (eds) *Birds and Pastoral Agriculture in Europe*. Joint Nature Conservation Committee, Peterborough.
Watson, J. (1992) Golden Eagle *Aquila chrysaetos* breeding success and afforestation in Argyll. *Bird Study*, 39: 203–206.
Watson, J. (1997) *The Golden Eagle*. Poyser. London.
Watson, J. and Dennis, R. H. (1992) Nest site selection by Golden Eagles *Aquila chrysaetos* in Scotland. *British Birds*, 85: 469–481.
Watson, J., Rae, S. R. and Stillman, R. (1992) Nesting density and breeding success of Golden Eagles *Aquila chrysaetos* in relation to food supply in Scotland. *Journal of Animal Ecology*, 61: 543–550.
Watson, J., Leitch, A. F. and Rae, S. R. (1993) The diet of Golden Eagles *Aquila chrysaetos* in Scotland. *Ibis*, 135: 387–393.
Watt, A. S. (1955) Bracken versus heather, a study in plant sociology. *Journal of Ecology*, 43: 490–506.
Watt, A. S. (1976) The ecological status of bracken. *Botanical Journal of the Linnean Society*, 73: 217–39.
Watt, A. S. (1979) A note on aeration and aerenchyma in the rhizome of bracken (*Pteridium acquilinum* (L) Kuhn var. *aquilinum*). *New Phytologist*, 82: 769–76.
Weir, T. (1993) Opening to the Conference. Pp. 11–14 in A. Watson and J. Conroy (eds) *The Cairngorms: Planning Ahead*. Kincardine and Deeside District Council, Stoneavon.
Welch, D., Scott, D. and Staines, B. W. (1992) Study on effects of wintering red deer on heather moorland: report of work done April 1992–November 1992. *Report to Scottish Natural Heritage*, 07/91/F2A/218.
Whitehead, G. K. (1964) *The Deer of Great Britain and Ireland*. Routledge and Kegan Paul, London.
Williams, A. G., Kent, M. and Ternan, J. L. (1987) Quantity and quality of bracken throughfall, stemflow and litterflow in a Dartmoor catchment. *Journal of Applied Ecology*, 24: 217–230.
Williams, G. H. (1980) *Bracken Control: a Review of Progress, 1974–79*. (Research and Development Publication no. 12). West of Scotland Agricultural College, Auchincruive,
Williams, G. H. and Foley, A. (1976) Seasonal

variations in the carbohydrate content of bracken. *Botanical Journal of the Linnean Society*, 73: 87–94.

Wolfe, A., Whelan, J. and Hayden T. J. (1996) The diet of the mountain hare (*Lepus timidus hibernicus*) on coastal grassland. *Journal of Zoology*, 240: 804–810.

Woolgrove, C. E. and Woodin, S. J. (1996) Current and historical relationships between the tissue nitrogen content of a snowbed bryophyte and nitrogenous air pollution. *Environmental Pollution*, 91: 283–288.

Wright, R. F., Crosby, B. J., Ferrier, R. C., Jenkins, A., Bulger, A. J. and Harrison, R. (1994) Changes in acidification of lochs of Galloway, south western Scotland 1979–1988. The Magic model used to evaluate the role of afforestation, calculate critical loads and predict fish status. Journal of Hydrology, 161: 257–289.

Yalden, D. W. (1981) Loss of grouse. Pp. 200–203 in J. Phillips, D. Yalden and J. Tallis (eds) *Peak District Moorland Erosion Study. Phase 1 Report*. Peak Park Planning Board, Bakewell.

Yalden, D. W. and Yalden, P. E. (1989) The sensitivity of breeding golden plovers, *Pluvialis apracius*, to human intruders. *Bird Study*, 36: 49–55.

Yalden, D. W. and Yalden, P. E. (1990) Recreational disturbance of breeding golden plovers, *Pluvialis apricarius*. *Biological Conservation*, 51: 243–262.

Zehetmayr, J. W. L. (1987) Influences shaping development in plantation forestry in Wales. Pp. 5–8 in J. E. G. Good (ed.) *Environmental Aspects of Plantation Forestry in Wales*. Institute of Terrestrial Ecology. Grange over Sands.

SUBJECT INDEX

•

SPECIES INDEX